D0087681

MOLDS, MOLECULES, AND METAZOA

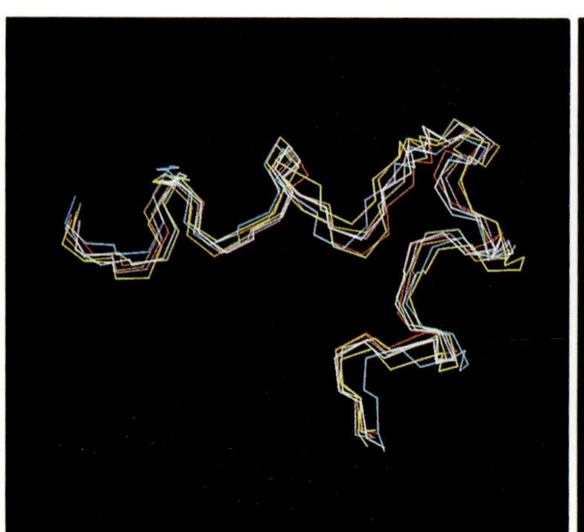
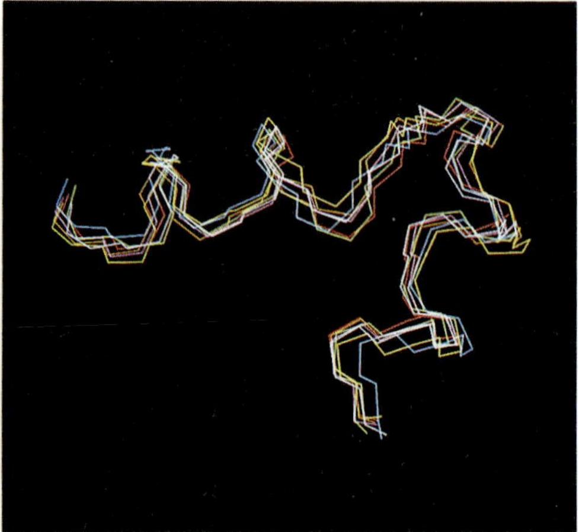

Stereo view of the superposition of the helix-turn-helix motif (residues 30–50) in the homeodomain (blue) with the helix-turn-helix motifs in phage 434 repressor (red), lambda repressor (violet), 434 Cro protein (brown), Trp repressor (yellow), CAP (green), and Lac repressor (white). The remarkable structural conservation of the helix-turn-helix motif may reflect an equally remarkable evolutionary conservation of a protein domain. (From Gehring et al. 1991)

MOLDS, MOLECULES, AND METAZOA

GROWING POINTS IN EVOLUTIONARY BIOLOGY

EDITED BY

Peter R. Grant AND **Henry S. Horn**

PRINCETON UNIVERSITY PRESS

Published by Princeton University Press, 41 William Street,
Princeton, New Jersey 08540
In the United Kingdom: Princeton University Press, Oxford

Library of Congress Cataloging-in-Publication Data

Molds, molecules, and metazoa: growing points in evolutionary biology
/ edited by Peter R. Grant and Henry S. Horn.
p. cm.
Based on a symposium held at Princeton University on Oct. 13, 1990.
Includes bibliographical references and index.
ISBN 0-691-08768-7
1. Evolution (Biology)—Congresses. I. Grant, Peter R., 1936–
II. Horn, Henry S., 1941–
QH359.M64 1992
575—dc20 91-30846

This book has been composed in Linotron Times Roman

Princeton University Press books are printed on
acid-free paper, and meet the guidelines for permanence
and durability of the Committee on Production Guidelines
for Book Longevity of the Council
on Library Resources

Printed in the United States of America

10 9 8 7 6 5 4 3 2 1

Contents

Contributors

Graham Bell
Department of Biology, McGill University
1205 Docteur Penfield Avenue, Montreal, P.Q. H3A 1B1, Canada

John Tyler Bonner
Department of Ecology and Evolutionary Biology
Princeton University, Princeton, NJ 08544-1003, USA

Leo W. Buss
Department of Biology, Yale University
New Haven, CT 06511-7444, USA

Matthew Dick
Department of Biology, Yale University
New Haven, CT 06511-7444, USA

Peter R. Grant
Department of Ecology and Evolutionary Biology
Princeton University, Princeton, NJ 08544-1003, USA

Marc Kirschner
Department of Biochemistry, University of California
San Francisco, CA 94143-0448, USA

Martin Kreitman
Department of Ecology and Evolution, University of Chicago
1103 East 57th Street, Chicago, IL 60637, USA

James W. Valentine
Department of Integrative Biology, Museum of Paleontology
University of California, Berkeley, CA 94720, USA

Mary Jane West-Eberhard
Escuela de Biología, Universidad de Costa Rica
Ciudad Universitaria, Costa Rica

Preface

About fifty years ago the Modern Synthesis of evolutionary fact and theory was born. The exact birthdate is unknown, and the birth certificate has been lost, but 1940 is as good a starting point as any. The synthesis did not satisfy everyone, least of all representatives of the most disparate and least integrated disciplines—genetics and paleontology. A committee on Common Problems of Genetics, Paleontology, and Systematics was established in 1943 to do a better job of integration. The product of their labors was an international Conference on Genetics, Paleontology, and Evolution held at Princeton University in 1947. The proceedings were published as a book (Jepsen, Mayr, and Simpson 1949).

Knowledge and understanding of evolution have been transformed in the last fifty years, and the pace of that change has been increasing due mainly to the extraordinary advances made by modern molecular biology. To keep the transformation in perspecive as it has occurred, to measure progress, expose problems, and identify the most promising lines of inquiry for the future, symposia have been held periodically, and books have been published from them. The present volume is in this tradition. However, rather than attempt, like many of our predecessors, to integrate current knowledge from all of the relevant disciplines—paleontology, systematics, genetics, ecology, behavior, neurobiology, physiology, and molecular, cellular, and developmental biology—we have chosen six of the major disciplines and asked, Where are they heading?

The symposium that forms the basis of this book was organized by Mal Steinberg and the editors, and held at Princeton University on October 13, 1990, to honor an evolutionary biologist of extraordinary breadth: John Tyler Bonner. By a curious coincidence, his career at Princeton exactly spans the interval between the 1947 conference and the present one. In 1990 he retired after a distinguished career as scholar and teacher—an educator in the broad sense—and administrator; for fourteen years he was chairman of the Department of Biology.

The theme of the symposium was simply evolution, because this was the main theme that runs through his long and broad research career, from his pioneering work on the mechanisms involved in the aggregation behavior of slime molds to the phenomenon of size scaling across all taxa. For example, he has written *The Evolution of Development* (1958), *The Evolution of Culture in Animals* (1980), and *The Evolution of Complexity by Means of Natural Selection* (1988), to name only three of his twelve books. In his words, "Natural selection, and our whole conception of evolution remains the most encompassing and the most useful theoretical framework that exists in biology" (*Size and Cycle*, 1965, p. 6).

It was tempting to look backwards over the last fifty years, or the last forty-three years since the previous Princeton meeting, to survey the transformation of the Modern Synthesis into what is sometimes called neo-Darwinism. In honoring John Bonner, however, we hoped to capture the spirit of his research with a forward orientation rather than a retrospective view. Six speakers were invited to contribute to the symposium by charting the future course at six major growing points in the study of evolution. And to have it both ways, to combine retrospect with prospect, we invited John Bonner to contribute the opening chapter of this book, and to provide a personal account of the major events of the last fifty years, which bring us to the excitement of the present.

Following the opening chapter the six main areas are arranged, appropriately, in a linear array of different scales. We start with evolution over geological time and finish with molecular genetics—from macroevolution to the ultimate in microevolution. To use an analogy, it is as if we were to take the subject of evolution, place it under a microscope, and view it under six different objective lenses with different powers of magnification. The first, with lowest power, has the widest range. We start with a paleontologist's view of current problems and promising directions of research in evolution, proceed to the ecology of populations and communities, and from there, behavior and related properties of individuals. Racking down the microscope we view developmental processes within individuals, then processes at the cellular level, and finally evolution at the molecular level.

Each contributor, while an expert in his or her own field, is not restricted to that field. At all levels, but particularly at the cellular and developmental levels, interconnections will be manifest. This, too, is appropriate, for John Bonner attempted to connect molecular, developmental, and evolutionary events into a single integrated scheme, his own personal modern synthesis, finding the life cycle and its constituents to be an appropriate organizing device (Bonner 1965).

MOLDS, MOLECULES, AND METAZOA

1

Evolution and the Rest of Biology: The Past

JOHN TYLER BONNER

My passion for biology spans over fifty years, and during the course of those fifty-odd years two things have happened: one to me and one to the world. In my own case, as the years progressed, I have become increasingly obsessed with the idea that evolution by natural selection is the most useful, the most important, the most all-enveloping concept in all of biology. That has been my own evolution. In the case of the world at large, there have been the most staggering advances and rapid changes of monumental importance in biology during this period. Here I would like to trace, with a very light brush, what those changes have been. My purpose is to bring the reader rapidly across the terrain of this half century, right to the edge of the cliff, to the beginning of the unknown void of the next fifty years. I do this keeping in mind that we are not just talking about the advances in any particular field, but how those advances are held together by the glue of evolution. The chapters in this volume show what some gifted young scholars, preeminent in their respective fields, can see as they peer over the precipice and look into the future.

Paleontology

In his *On the Origin of Species*, Darwin argued that one of the reasons it is difficult to trace the lineage of species through the fossil record is that the record is so imperfect—there are so many gaps. What he said then could be justifiably said today, but the difference in our knowledge is enormous. Today, the total number of fossils known is vast compared to what was available in the middle of the last century. Furthermore, with our methods of dating fossils by combining the geological knowledge of the sequence of strata with methods using radioisotopes, we have in recent years developed a remarkably accurate time course for our fossil sequence. Therefore, we have done much to fill in the gaps deplored by Darwin but, like the pursuit of fractals, we find that with increased magnification and precision the gaps may be smaller, but they are still there, and, if anything, they are more numerous than before.

There have been some very exciting discoveries in the never-ending search for new fossils. For instance, there was the remarkable finding of fossils of very early, primordial bacteria. There was also the discovery (and, more recently, the interpretation) of the ancient fossils of the Burgess Shale which reveal the soft parts—not just the skeletal armor—of a wide and fascinating variety of forms of invertebrates. These discoveries, and others like them, have provided for a steady accumulation of new information, new facts about the plants and animals of the past.

Significant as these discoveries are, the great advances came from making useful generalizations from these facts; in a sense, they are the grist for the all-important mill of insightful models. Certainly, even in the last century, there have been attempts to make broad generalizations about trends in fossil sequences, in fossil history. There was Cope's rule of trends in size increase and other such generalizations which are still of use to us today. The revolution, however, came from the work of the paleontologist G. G. Simpson. He was the first to look into the matter of rates of evolution. He was able to calculate the rates of origin of new species, new genera, new orders, and so forth, at different periods of earth history, as well as rates of extinction, and in this way obtain a picture of what he called in the titles of one of his influential books "the major features of evolution" and, in another, "tempo and mode in evolution." He could use these quantitative methods for the study of rates of change because the accumulation of facts about fossils had reached a sufficient level of detail, and with these facts he was able to weave together some generalizations of major significance.

It is this approach that has dominated paleontology in recent years. Sometimes it has produced generalizations that are in complete harmony with population genetics and neo-Darwinism. This certainly was the view of G. G. Simpson. Recently there has been a heated debate as to whether evolutionary change is gradual or involves great bursts followed by periods of stasis (the so-called punctuated equilibria model). Sometimes there has been more heat than light generated in this debate and, as is so often the case, both sides are no doubt correct. In any event, there is nothing to seriously shake the foundation of evolution by natural selection, although there could be special features of the speed of change in speciation that we do not yet fully understand. This is the edge of the precipice of interpreting the fossil record: we have new tools and new methods. The future should be exhilarating.

Ecology

In 1941, just after I received my B.S. degree, I spent a good part of one summer in Barro Colorado Island. It was a great awakening for me. In the first place, callow as I was, I was absolutely overwhelmed by the richness of the tropical rain forest. It was far beyond anything my imagination or reading had ever conjured up in my mind. The other was a week's visit there by Professor Victor Shelford, then an old man (or at least it seemed so to me at the time) who was well known for his ecology textbook. We had many conversations about biology and ecology in which I gained a clear impression that in his view one must not tamper with nature and experiments were bound to be misleading because they were not "natural." What the ecologist should do is try to appreciate the great complexity of the biosphere and record it as accurately and as systematically as possible. Since I was a confirmed experimentalist, I spent some time vigorously arguing with him; but despite our many disagreements (which included politics and Franklin Delano Roosevelt), I did see his point about complexity and the extraordinary interrelations and interdependence among all the parts of an ecosystem.

The great change came in the 1950s and 1960s, when G. E. Hutchinson and, more especially, his student Robert MacArthur began to apply mathematical models in an attempt to make simplified generalizations about the great complexity of ecological systems. As we all know, MacArthur was remarkably successful in this enterprise and did indeed spawn a revolution. Testimony to its success was the great cry of resentment from the traditional ecologists of the older school, who felt that the important aspect of an ecosystem was its complexity; anything that tried to simplify it was working in the wrong direction. But the MacArthurian insights were powerful and they carried the day. Furthermore, they greatly helped to link ecology with evolution, which he and G. E. Hutchinson and others did so much to encourage.

Since that golden age there has been a refinement of the models. Many of the original ones have even been supplanted, as MacArthur predicted they would be. It has also been a period of exponential proliferation of models; the approach has been so fruitful that it has seeped into every corner of ecology and across the boundary into behavior and evolution. There is no doubt that this has been one of the most successful cases so far in biology, where mathematics has helped to give insight into complex mechanisms and interactions of many kinds. We can easily imagine that the same will be true for other fields of biology, but for ecology it has been a fertile period for the use of mathematics as a way to shed light on a difficult ecological problem. The days of "thick description" are happily gone. This is where we are now, at the edge of the precipice that leads to the future. I cannot imagine that the next steps will be more of the same; perhaps we are ready for a new insight or a new approach.

Behavior

There always seemed to be some confusion in Darwin's mind about what he called instinct. It no doubt stemmed from his inability to account for the mechanism of inheritance. In a number of passages in his writings he makes it clear that he sees a difference in how instincts might be inherited, in contrast to morphological changes, but he is never explicit on what these changes might be. Yet he frequently seems to imply that instincts are inherited in a rather Lamarckian fashion.

Well into this century the problem occupied the minds of many and led to a great dichotomy: those who became enamored with the idea that behaviors (instincts) were inherited, of which the eugenicists were the most notorious in that they were considered synonymous with bigots, and those who said that the environment and learning were the sole determinants of behavior. This became the rather sad and pointless nature-nurture debate. When I took elementary biology, the Jukes and the Kalikaks, families that spawned an exceptionally large number of criminals over a number of generations, were proof that delinquency was indeed inherited. Hardly ten years later this "fact" was expunged from all biology textbooks, only to reappear in sociology texts as evidence that the environment had a profound effect on behavior. In fact, the word "instinct" itself became disreputable, because how could one ever know the genetic component in any behavioral act?

The revolution came with the work of Konrad Lorenz and Niko Tinbergen and others in the birth of ethology. They made it possible, by the carefully controlled study of animals in nature, to be able to distinguish between instinctive and learned behaviors, and "instinct" was able to regain its former respectable reputation. In this ethological revival there was an immediate concern with how instinctive behaviors were controlled by natural selection, and a convincing case was made to show that the behaviors, which were automatic and clearly inherited, did endow fitness, and therefore reproductive success, to those individuals that inherited the behaviors.

The culmination of this approach came in the enormously important inspiration of W. D. Hamilton in the 1960s—that the key to understanding the evolution of social behavior was the genetic relatedness of the individuals involved. Instinctively helping kin could appear to be altruistic, but in fact there was also a selfish component to the behavior, so that by such apparent altruism one made it more likely that the genes one shared with close relatives would be passed on to the next generation. This raised a great interest in animal societies, which came to a peak in the publication by E. O. Wilson in 1975 of his important book, *Sociobiology*. Unfortunately it was met not only with acclaim by biologists in general, but with vitriolic abuse from Marxists who used it to revive the moribund nature-nurture problem and all its sinister possibilities for eugenics. This attack had nothing to do with the main body of the work, but only considered the implications for human sociobiology, which was not part of Wilson's central agenda, nor is it to this day a subject tractable for sensible study. Perhaps the only good that has come out of the human aspects of sociobiology is to stimulate the beginning of a greater awareness of the interactions between cultural and genetic inheritance, a field that is still in its infancy.

Much of the recent study of interactions in animal societies has, quite obviously, centered around means of communication among individuals. Some of the most significant studies stem from the social insects, for they have so many different kinds of social groupings, and some of them are so elaborate. For example, Karl von Frisch showed that scout bees could tell the other foraging worker bees both the direction and the distance of a source of nectar, a discovery so surprising that few believed the story when they first heard the rudiments of it. And it was also the hymenoptera that gave W. D. Hamilton the idea that has come to be known as kin selection.

The fact that behavior and evolution are intimately entwined has never been more evident than it is now. For many reasons—most of them made possible only in the last fifty years—we see the problem far more clearly than Darwin could. This is where we stand on the cliff looking out into the future.

Development

In the case of development, there are two very distinct traditions. One is the explanation of how plants and animals develop: this is the "mechanics of development'' of Wilhelm Roux in the latter part of the last century and the ''casual embryology'' of Albert Dalcq in the early part of this century. The other is the relation of development to evolution, a tradition that goes back at least to Karl Ernst von Baer in the early 1800s, and later, as the nineteenth century progressed, many others came to make important contributions to the subject. They include Charles Darwin, August Weismann, and Ernst Haeckel, and early in the twentieth century there were Walter Garstang and Gavin de Beer. (The roots go even farther back, as Stephen J. Gould shows admirably in his book, *Ontogeny and Phylogeny*.) Let me discuss briefly both approaches—the how and the why of development—and then show how they seem to have come closer together in recent years.

Twenty or thirty years ago it was often said that the golden age of developmental biology peaked early in the twentieth century, with the discovery by Hans Spemann of the ''organizer'' in the amphibian embryo, which demonstrated that one part of an embryo could send chemical messages to another and in this way call forth or induce new structures in undifferentiated tissues. It was the beginning of a great search for what are now called ''morphogens,'' chemical signals which coordinate and orchestrate the development of an organism. In plants, the discovery of auxin and other growth hormones has been of major importance, and although the chemical identification of the morphogens in animal embryos has been slow, recently we have seen major new discoveries, such as the role of retinoic acid in limb development in vertebrates. The other modern feature of this offshoot from Spemann's work has been the use of mathematical modeling. It was given some initial sparks in the work of Nicolas Rashevsky, but generally the broad concepts of reaction-diffusion mathematics and how they might apply to development are attributed to Alan Turing in his famous paper of 1952. From all this we can see that while Spemann's work opened up a wide avenue of research, that avenue has been productive more or less continuously for many years.

The more recent development, and one of at least equal, and probably greater, importance, is the rise of developmental genetics. It has been known for a very long time that much of the information for development is stored in the genes of a fertilized egg, but what has been unclear until recently is how the genes impart their information to the embryo. The origins of this approach come partly from classical genetics and partly from the great rise of molecular genetics following the megadiscovery by James Watson and Francis Crick of the structure of DNA, which led to our understanding of how the genetic code works and how the genes designate the synthesis of specific proteins.

With these tools and associated clever experimental methods, it is now possible to identify the structure of specific genes, the structure of the proteins they produce, where the proteins are located in the embryo, and even where they are located in the cells of the embryo. With this incredibly powerful armory of techniques it is now possible to trace in great detail the chemical course of events in the development of an organism. This work is being done with notable success in the development of the fruit fly *Drosophila* and the nematode *Caenorhabditis*, which are being intensively studied in many laboratories. They make a particularly happy choice because the early development of *Drosophila* is to a considerable extent regulative, and the gene products are in many cases themselves morphogens that signal from one region of the embryo to another. In contrast, the development of *Caenorhabditis* is relatively mosaic and the cell lineages are fixed. Yet in both there are instances where individual cells can signal to their immediate neighbors, and in some cases we know the structure of genes and the structure of the signal proteins as well as that of the receptor protein. The future of "how" development takes place could not look brighter than it does at the moment.

It would be wrong not to add that no one thinks development can be explained solely in terms of gene products. One must include the properties of those products. They might be adhesion molecules, or molecules that produce basement membranes, or molecules that can contract, and these properties will play an important role in the final outcome of the development. The genetics of development must never be considered in the abstract: the genes may pull the strings of the marionette, the cells to which they are attached; but the properties of the cells and the gene products play a major role in bringing the organism, that is, the marionette, to life.

If we turn to how development plays a role in evolution, the principle theme which flourished especially in the latter part of the nineteenth century and the twentieth was heterochony. By changing the timing of development of parts of the embryo, it is possible to modify adult organisms in major ways. A classical example is neoteny in human beings. In our bodily adult features we closely resemble the juvenile stages of our primate ancestors. Compared to the apes, we have a greatly extended period of infancy and childhood development. This not only means that as adults we have the facial structure and hairlessness of juvenile or fetal apes, but also that our extended period of development is an important factor, allowing for the increase in growth of our brain relative to our body size.

Although there is continuing awareness of the role of heterochrony in evolutionary change, this is no longer the most important link between development and evolution. Rather, the question revolves around why does one have development at all, and since one does, how is it affected by evolution? The answer to the first question lies in the simple idea that it was clearly adaptive for some organisms to evolve large phenotypes to carry the germ cells safely and effectively to the next generation. The second question is more subtle and raises the matter of what is selection acting upon. It is clearly not just the adult, as Garstang pointed out many years ago; the whole life cycle is subject to selection. In many ways, Weismann was the first to have some insight into this problem, and currently Leo Buss has shown that the whole matter can be framed in terms of levels of selection. Not only are the genes and the whole phenotype subject to selection, but in primitive multicellular organisms there can be a competition between cell lineages.

The fact that developmental genetics is dealing directly with genes and their products, and that evolution by natural selection ultimately involves the selection of genes, is the reason for my saying earlier that the two approaches, developmental mechanics and the evolution of development, are inevitably coming closer together. Whatever lies across the chasm in the future will undoubtedly involve both aspects of development.

Cell Biology

Fifty years ago a good portion of everything to be said about the biology of the cell was in the third edition of E. B. Wilson's *The Cell in Development and Heredity*. This remarkable book, which had great influence, was a rich summary of the rise of cytology, the understanding of the chromosomes in heredity, and the cellular events that took place, especially in the early development of animal embryos. However, even fifty years ago, an enormous change in cell biology was already beginning to take place. This was due to the rise of biochemistry and the beginning of our appreciation of the vast number of chemicals within a cell, and how those chemicals were the mediators of metabolism, locomotion, growth, and all the other activities of cells. The progress of our knowledge of cell biochemistry has increased steadily and rapidly, and even today the pace has not slackened. We are constantly reading about new discoveries of major importance and significance.

We have come an extraordinarily long way in these five decades. We now are beginning to have a deep understanding of the chemical structure of membranes, of cells, the organelles within the cells, and the nucleus. We know where transcription of the DNA to RNA takes place, and we know how and where the RNA translates the code into specific proteins. We know how the proteins are transported into membranes and across membranes. We have discovered many new structural proteins and have found out how they assemble into microtubules and other structural elements within the cell, such as the spindle, which plays a major role in mitosis. We know much about the molecules which are responsible for cells sticking together in multicellular animals and which play so important a role in development. There is much more besides: one need only look at a current text in cell biology, such as the most recent edition of the pioneering *Molecular Biology of the Cell* by Alberts et al. But on many subjects the book may already be out of date, for the field is progressing at such a dizzy pace that important new advances are appearing with relentless regularity and speed.

Part of the great progress is due to the marvels of new techniques. To mention a few, it is now possible, as I said earlier, not only to identify key proteins, but to learn their complete amino acid sequence as well as the structure of the DNA that gave rise to it. Furthermore there are methods involving the use of specific antibodies attached to fluorescent labels so that the exact location of the substances can be found within the cell. One can, for instance, block a gene by adding some DNA which is in the reverse orientation of a key gene, and in this way prevent the synthesis of a particular protein and then study the result of such loss on the activity of the cell, or on the development of an embryo.

One final example. One of the subjects of particular interest to cell biologists, going back to E. B. Wilson and before, is the mechanism of cell division and mitosis. We hardly understand it all today, but our knowledge is extensive due to the discoveries of many workers in many laboratories. We know that there are key chemicals which accumulate and are destroyed and that the cell division cycle is regulated by these substances. We are beginning to have a picture of how the spindle microtubules actually separate the chromosomes to the two daughter cells. And, of course, there is much more.

To me a most impressive aspect of all these advances is that it leaves one with the realization that a tiny cell is an extraordinarily complex chemical factory. The number of key elements or chemical steps involved in any one function seems large, and as new discoveries are made, it gets larger. Secretion, movement, mitosis and cleavage, protein synthesis, energy conversion in the mitochondria—all involve so many steps that it is increasingly hard to imagine all that taking place in such a minute space.

To turn now to the relation of cell biology and evolution, there is far less to say. There is, of course, the question of the origin of the cell, and how prokaryotic cells evolved into eukaryotic ones, and both problems have stimulated much discussion and interesting speculation. Perhaps the more relevant point centers around the fact that once cells were invented they seemed to have remained pretty much the same. In the case of eukaryotic cells, this appears to have been an exceptionally successful structure because it became the building block of all higher plants and animals.

At least on the surface, there seems to have been an enormous conservatism in cell structure and function, and the cells of all eukaryotic organisms have a very similar mitosis, meiosis, and fertilization process, as well as similar means of movement and of processing energy. Here lies the interesting point that is unresolved: it is not exactly the same in detail in all organisms. There are variations, albeit minor ones, in different groups of animals and plants, and those variations could be of major interest. Especially interesting are cases where similar chemicals, or similar chemical pathways, are used for different purposes in different organisms. Whether it be DNA or the direct or indirect products of DNA, it may be easier to modify them for a new function than to invest a whole new pathway.

Finally, I should say that inevitably it will be impossible to separate the relation of cell biology to developmental biology from their relation to evolution. It is becoming increasingly clear that the study of development is presently being carried out simultaneously on the level of molecular genetics and cell biology. Because of the power of these two approaches, they already dominate the study of development and undoubtedly will do so into the forseeable future. This means that strengthening the relation of evolution to development will involve a much richer and more encompassing meaning of the conception of development.

Molecular Genetics

At the moment it seems hard to imagine, but during the early part of my career the chemical composition of the genes was unknown. There were many opinions, and even some very strong evidence from Avery and his colleagues, but the sunburst occurred when, in 1953 Watson and Crick published their famous paper. It was not so much that they proved genes were made of DNA; they showed how the DNA could do the work of making complementary templates. Once this revelation was out it was only a matter of time to crack their code and show how base triplets of DNA coded for specific amino acids. A mere four bases of DNA could code for twenty amino acids. Furthermore, all the chemical machinery, the enzymes, and the structure necessary to produce this result were soon to be discovered. Not long afterwards the essence of the story could be found in elementary textbooks of general biology. It was as though a cyclone had hit biology, and the rate at which the field has continued to develop, and continues this very minute, is a miracle. There is a vast array of molecular biologists, often our very brightest students, who are pressing the pace of this wonderful progress.

The contributions made by molecular biology to solving problems of evolution have been extremely modest by comparison. Again there have been those who have speculated on the origin of the code and DNA and RNA, but as interesting as these conjectures might be, they do not satisfy me any more than do the speculations of cosmologists on the origin of the universe. I suppose it is simply too hypothetical; the clever ideas do loop-the-loops around the facts, and not always with constraint.

A far more practical relation between molecular biology and evolution has come from the comparison of specific proteins or the DNA of different organisms to determine how closely related they might be, and in this fashion to provide another way to determine the phylogenetic relation between any two organisms. The first work was done on proteins, and one could compare, for instance, the number of amino acid differences in a protein such as hemoglobin in different species of vertebrates. If one assumes that the rate of mutation, and therefore amino acid change, is approximately constant on average over geological time, then one can estimate when two kinds of vertebrates diverged from one another. To put this in terms of years, rather than in relative terms, it is necessary to correlate the molecular information with our paleontological knowledge, which has given us some moderately accurate dating by measuring the ratios of radioisotopes in the fossil rocks. It is clear that some proteins change far more rapidly than others, and some are present in most organisms (such as cytochromes), while others are largely confined to one group (as is hemoglobin in vertebrates). The critical assumption of constancy in the rate of amino acid substitution over geological time turns out to be rather uncertain.

There are a number of other molecular methods that are currently in use to measure phylogenetic distance or closeness. For instance, it is possible to mix the DNA strands of two organisms together to see, by the degree to which they stick together under rather special conditions, how similar the species are genetically. This so-called DNA hybridization has both advantages and technical problems associated with it, so that it is also fraught with controversy. Another approach has been to study some specific part of the DNA, such as that which codes for the ribosome. Finally, there are strong advocates of the use of mitochondrial DNA to compare species because, among other things, it evolves faster than nuclear DNA. Each of these methods has advantages and disadvantages, so not only do practitioners argue among themselves as to the reliability of their methods, but there are further squabbles when morphological changes seem to be more rapid than the molecular changes. No doubt the problems of these methods will be slowly ironed out, and already we have seen some important and revealing insights into animal and plant phylogeny.

As interesting and as useful as molecular phylogeny has been and will be, that in itself is hardly more than a guide or stepping-stone to the future. It is my ever-optimistic hope that molecular biology will lead to much deeper insights than who is more closely related to whom. And indeed this has already begun in the study of variation itself, the basic grist for the mill of natural selection.

NO doubt all fields have changed radically in the last fifty years, but it is hard to believe any have done so with quite the vim and vigor of biology. As I said in the beginning, what impresses me most about our new knowledge is that every corner of that knowledge is directly related to evolution by means of natural selection. In a few cases we can show directly how selection has produced changes; in many others selection is inferred. In order for evolution to occur by selection one must have genetically inherited variation. Variation can be selectively neutral (and give us molecular clocks), but as soon as it imparts an advantage it will be pushed forward by reproductive success. This means that paleontology gives us the evidence that natural selection has been operating on earth for billions of years. It means that selection occurs in an environment and therefore ecology is the study of the setting in which selection operates. It means that animal behavior arose by natural selection, and once invented it allowed evolution to occur in novel ways. Selection operates at all points in entire life cycles. It operates on variation among organisms as wholes as well as on the minute machinery of their cells and the molecules that make up those cells.

None of these subjects—paleontology, ecology, behavior, development, cell biology, and molecular genetics—has any lasting meaning unless they are threaded together by the evolution that brought them into being.

2

Lessons from the History of Life

JAMES W. VALENTINE

Santayana's famous aphorism about being doomed to repeat history when in ignorance of it is not without relevance to the history of life. Were we to understand the history of life, we might be able to avoid the creation, through human activities, of disasters and catastrophes such as have been visited naturally upon the biosphere in the past. While studies of the living biota by systematists, ecologists, and evolutionists are absolutely necessary for conserving the biosphere in the face of changes driven by human population, they may not be sufficient by themselves to permit the depth of understanding necessary for this purpose. If they are not, it is probably because the long-term consequences of some biological processes and events are not evident, and perhaps some cannot be inferred in principle, from studies prosecuted over the scale of years in which biological scientists must work. Although the fossil record is not a very good place to study processes, it is a much better place to learn their consequences, especially in the long term.

The field of invertebrate paleontology, which is my own, was in early decades of this century essentially a field of geology, at least in America. It has become increasingly biological since the Second World War, beginning with paleoecology and continuing with macroevolution and other disciplines. I believe that this trend towards a more biological paleontology will continue. Here I review three areas of marine invertebrate paleobiological study in which neontological work must be involved if important progress is to be made. Although they do not bear on conservation, these are areas for which history is clearly important to the science of biology.

The Early Metazoan Radiations

The first paleobiological area concerns the origin and establishment of major animal groups such as phyla. During the last few decades, knowledge of the fossil record of early animals has increased enormously (see, for example, Glaessner 1984 and Fedonkin 1985a,b), but the record has not seemed to make any special sense with prevailing scenarios of phylogeny. With the advent of molecular work on phylogenies and on developmental genetics, however, these scenarios are being revised, and it now seems possible to rationalize them with the fossil data. The general sequence of major biological events during the origin and radiation of major groups is summarized in table 2.1. The earliest fossils likely to be bilaterians, of late Precambrian (Vendian) age, are traces of creeping or plowing worms that left surface trails, and perhaps shallow horizontal burrows or furrows, but that are not identified with any particular body fossils. Macrofaunal burrows that penetrate deeply or vertically into sediment are not certainly known until either the very latest Vendian or the earliest Cambrian. Associated with the horizontal traces are enigmatic body fossils. Many of these organisms, which are repre-

Table 2.1. Sequence of appearance of faunal elements during the Precambrian-Cambrian transition, when the body plans of most metazoan phyla appear to have originated.

	Lower Cambrian
Botomian	15 ± metazoan orders make first appearance.
Atdabanian	29 ± metazoan orders make first appearance. Earliest body fossils of both durably skeletonized and soft-bodied trilobites; first echinoderms; soft-bodied fossils such as *Microdictyon*. Vertical burrows larger, more common, bioturbation deepens.
Tommotian	22 ± metazoan orders make first appearance. Earliest mollusks and brachiopods. Trace fossils diversify and increase in abundance, earliest? arthropod traces, vertical burrows still small but more common, bioturbation shallow.
	Late Precambrian
Late Vendian	24 ± metazoan orders make first appearance. Earliest skeletons of problematica; very small vertical burrows. Vermiform, segmented soft-bodied fossils of uncertain affinities and vaguely cnidarian fossils, termed "Vendozoa" (Seilacher, 1989), occur.
Early Vendian	Extensive (Varangian) glaciation. Only prokaryotes and single-celled eukaryotes are certainly known from fossil evidence.

NOTES: Both skeletal and soft-bodied fossils are termed "body fossils"; traces are signs of animal activity but not the remains of the animals themselves. Base of Cambrian is near 530 million years ago. (Ordinal number estimates from Valentine et al., in press.)

sented by impressions, appear to have been cnidarian grade, resembling medusae or sea pens. Some of them may represent living orders, but most are so distinctive that they are placed in orders or even classes of their own (as *Albumares* and other triradiate medusoid forms; figure 2.1).

The cnidarianlike fossils cannot be responsible for the trails. However, possible bilaterian body fossils are represented by such Vendian forms as the vendomiids and, especially, the sprigginids (figure 2.2). These fossils can be interpreted as lightly scleritized, hemocoelic, segmented creeping worms (Valentine 1989a). Some of the horizontal trails may have been made by such forms. Arthropodlike traces are also reported from the Vendian (Jenkins et al. 1983) but have not yet been illustrated or described in detail; no Vendian fossils with arthropodlike appendages are known.

Interpretation of the puzzling Vendian body fossils has ranged to extremes. One group of workers (epitomized by Glaessner 1984) has tended to place all or nearly all of these Vendian forms in living phyla, either because they share some particular feature such as segmentation, or because they possess an attribute of a hypothetical primitive stage of living phylum as envisaged in some phylogenetic scheme. At the opposite extreme, all Vendian body fossils have been assigned to a group of animals, termed the Vendozoa, that are suspected of not even being metazoan, and are claimed to display a unique constructional system (of

Figure 2.1. *Albumares brunsae* Fedonkin (X2.6) from the Vendian of the White Sea, USSR. Modern medusae have a basic fourfold symmetry, yet a number of Vendian medusoid forms such as this one have a threefold symmetry and have been distinguished as a new class, the Trilobozoa, by Fedonkin. (From Glaessner 1984.)

Figure 2.2. *Marywadea ovata* (Glaessner and Wade) (X2), a sprigginid from the Ediacarian beds of South Australia, roughly correlative with the Vendian of the USSR. This form has been interpreted as an annelid, as an arthropod, as representing a new phylum intermediate between those phyla, as a nonmetazoan ''Vendozoan,'' and as representing an intermediate grade between lower invertebrates and arthropods, the interpretation favored herein. (From Glaessner 1976.)

''pneus,'' hollow modules that are assembled, quilt fashion, to form the various architectures that are observed; see Seilacher 1984, 1989).

Just at the close of the Vendian, small mineralized skeletal elements of varying compositions make their appearance (figure 2.3) and become quite diverse in the earliest Cambrian stages—the ''small shelly'' faunas (Matthews and Missarzhevsky 1975; Rozanov 1986; Brasier 1989). These remains are chiefly sclerites—that is, dissociated elements of a scleritome or skeletal assemblage—and reconstructing the forms of the animals that bore them is exceedingly difficult. However, several reasonably complete scleritomes have been recovered, and some sclerites have been discovered in association with soft-bodied remains. Examples include minute phosphatic buttonlike structures (*Hadimopanella)* that are evidently dermal sclerites of groups of annulated worms, the paleoscolecids (Hinz et al. 1990); the spiny *Halkieria;* and the netlike sclerite, *Microdictyon*. This last form had been known for some time from disarticulated sclerites, and had attracted a variety of speculations as to its affinities and functions, but no one imagined that it was embedded above the tubular appendages of the implausible-looking invertebrate of figure 2.4, which turned up recently (Chen et al. 1989). The leading interpreter of early scleritomes admitted to feeling

Figure 2.3. A few small shelly fossils from the early Cambrian of the USSR (various enlargements between X25 and X40): 1, *Aldanella* (a mollusk?); 2, *Coleoloides* (affinity unknown); 3, *Chancelloria* (allied to sponges?); 4, 5, *Sachites* (probably related to *Halkieria*, a scleritome of which is known and may represent an extinct phylum); 6, 7, *Camenella* (affinity unknown); 8, 9, *Lapworthella* (affinity unknown); 10, *Fomitchella* (affinity unknown); 11, 12, *Protohertzina* (resembles spines of modern chaetognaths). (From Matthews and Missarzhevsky 1975.)

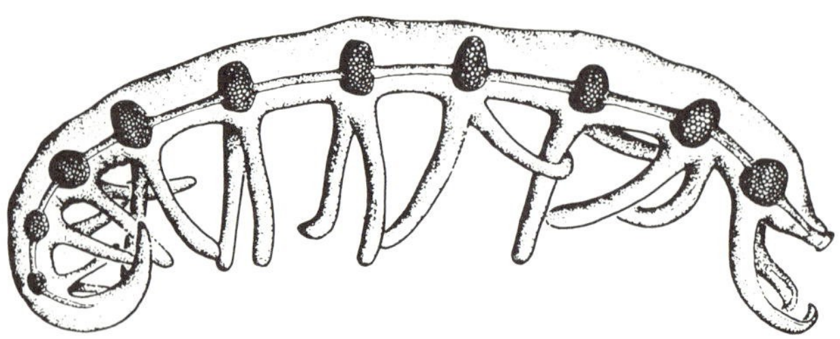

Figure 2.4. *Microdictyon sinicum* Chen, Hou, and Lu (approx. X3) from the Atdabanian of Yunnan, China. The oblong, netted structures overlying the appendages are phosphatic and have been known for some time; they were formally described in 1981 and the animal to which they belong was described in 1989. (After Chen et al. 1989.)

humbled by this discovery, a sentiment that is well known among those who deal with the early products of the metazoan radiations.

So far as can be told, few of these small shelly groups survived the Cambrian, although at least two living phyla, the brachiopods and mollusks, are first represented by small shelly fossils in the Tommotian, and trace fossils formed by arthropodlike appendages are certainly present then (table 2.1). In succeeding Cambrian stages additional living phyla appear, and among durably skeletonized bilaterians only the bryozoans are not found by the close of the Cambrian. Many of the extinct groups are represented by lightly skeletonized or unskeletonized forms that occur in the remarkable Middle Cambrian Burgess Shale of British Columbia (figure 2.5). Some of these fossil types are not known elsewhere, but a number of elements of this fauna are now known from the Lower Cambrian (Atdabanian) Chenjiang Formation of Hunnan, China, which yielded the *Microdictyon* animal. Despite the appearance of many living phyla, the Vendian-Cambrian faunas are not at all modern. About 155 orders can be characterized in the Vendian and Cambrian faunas, of which over 90% are extinct. A guess that I believe is conservative is that about 40% of the orders do not belong to living phyla (Valentine et al. 1991b). All of this high-level explosive diversification occurred at a time when species diversity was quite low by average Phanerozoic standards, although speciation rates were probably fairly high. One estimate is that during the height of the diversifications, in the Early Cambrian, about every fortieth species founded at least a class (Valentine and Erwin 1983).

What was going on? Neontological studies cannot help directly with

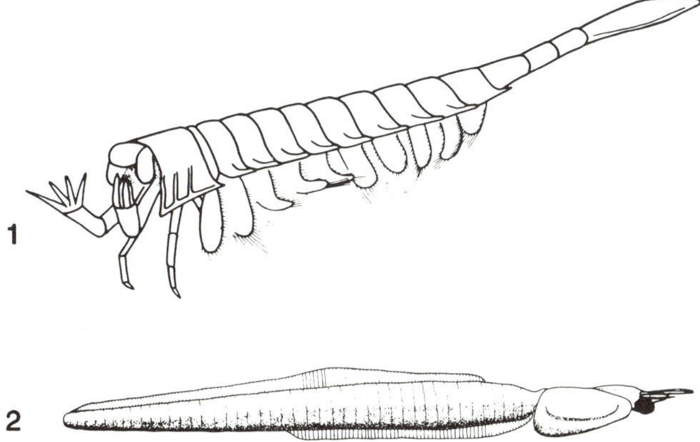

Figure 2.5. 1, *Yohoia tenuis* Walcott (approx. X9); and 2, *Nectocaris pteryx* Conway Morris (approx. X4.5), both from the mid-Middle Cambrian Burgess Shale of British Columbia, Canada. Although *Yohoia* is arthropodan in general aspect, it cannot be assigned to any living class, while *Nectocaris* cannot be assigned to any living phylum. Both appear to be segmented. (From Whittington 1974 and Conway Morris 1976.)

the many extinct clades, but they must be major contributors to the solution of this problem. First of all, they can provide evidence of the branching patterns that led to the surviving body plans. Molecular phylogenetic trees, as based on sequence similarities in 18S rRNA, provide some such evidence (Field et al. 1988; Ghiselin 1988; Lake 1990). In those molecular-based phylogenies, the order of branching varies somewhat according to the methods by which they are formed and the assortment of organisms included. An interesting feature of the trees is that arthropods appear to be less derived than annelids, and annelids less than the mollusks, brachiopods, sipunculids, and some other nonsegmented metazoans. These relations are subject to modification as new evidence is introduced, of course.

Furthermore, morphological studies have suggested that the branching between arthropod and annelid ancestors may have occurred before the body plan of either phylum was evolved, for the architectures of these plans are quite different. Arthropod architecture is based on the exoskeletal muscular support of a scleritized cuticle, with turgor and hydrostatic functions provided by a hemocoel; annelid architecture is based on a flexible cuticle with body-wall muscles antagonized by a hydrostatic coelom, commonly partitioned. The chief similarity between the body plans of these phyla is that they are both segmented.

Many phylogenies postulate that the arthropods, which appear to be the more complex, were derived from annelid ancestors (see Hyman 1940; Clark 1964). However, as the architectures and mechanics of their locomotory systems are so different, other workers have thought it more likely that they evolved independently (for example Manton 1977).

The arthropod architecture is clearly in aid of locomotion via appendages, and may have evolved from a segmented worm that moved via pedal waves (Valentine 1989a). Such a form may be represented by the sprigginids; the light scleritization that may account for their preservability (Seilacher 1984) was perhaps a coadaptation with segmentation for pedal locomotion. Clark (1964, 1979) made a strong case that the annelidan architecture originally evolved as an adaptation for efficient burrowing. Thus the horizontal Vendian trails are visualized as formed by descendants of acoelomates that, owing to size increases in some lineages, became, on one hand, increasingly complex, seriated, vascularized, and eventually scleritized, perhaps evolving originally on hardgrounds; and on the other hand, by some similarly complex forms with coelomic spaces and flexible, even extensible body walls adapted to creeping in soft sediments, a sort of advanced nemertine grade. In this scenario, at least one scleritized line developed segmentally arranged accessory gripping structures that evolved into jointed appendages and thus became arthropods. Another line or other lines became adapted to burrowing within confining sediment matrix, and exploited intramesodermal spaces to produce the eucoelomic segmented architecture of annelids. This adaptation evidently did not arise until the latest Precambrian and perhaps not until well into the Early Cambrian, when penetrating burrows begin to be common. Also from among the visualized plexus of segmented ancestors, other branches arose that may have included such nonannelidan, nonarthropodan forms as *Opabinia, Aysheaia, Nectocaris*, paleoscolecids (if they are not annelids), and others.

Some acoelomates and many nemertine-grade forms display seriation. It seems likely, as has been suggested a number of times, that the segmentation displayed by arthropods and annelids evolved from these patterns of seriation in lower invertebrates. Some nonsegmented protostomes also display seriation (such as chitons and monoplacophorans among the mollusks), which may be homologous with seriation in one or another of those lower invertebrates (Vagvolgyi 1967; Stasek 1972; Salvini-Plawen 1985). The branching that founded the ancestors of the mollusks and brachiopods must have occurred during the Vendian, as primitive groups of both these phyla appear in earliest Cambrian time.

It seems unlikely that annelids, as such, stood in the ancestry of these phyla, or of their close relatives.

Thus a combination of paleontological and neontological information permits us to postulate a Vendian radiation of acoelomate and perhaps primitive coelomate lineages that led eventually and independently to a number of higher invertebrate body plans (Valentine 1990a; see also Willmer 1990 and Bergstrom 1990). Some of these higher invertebrate stocks may have produced radiations of their own at the level of body plans that we call phyla; certainly many produced major modifications that are recognized as classes and orders today. However, the branchings that proceeded from the last common ancestors of many of the living phyla, and that may eventually be resolved by molecular techniques, may well have occurred among lower invertebrates. The rise of the higher invertebrate body plans appears to have been abrupt, to have been restricted to a relatively narrow window of geological time, to have involved many body plans that have not survived (those of which have descendants today were by no means dominant during the radiation), and indeed to have been responsible for the beginning of the Phanerozoic Eon.

Why this explosive radiation of body plans? Bonner's writings (1965, 1988) suggest an attractive possibility. Complexity of the sort found in higher invertebrates requires a number of organ systems, implying a variety of tissue types and in turn of numerous cell types. As Bonner has observed, cell type number may be the best single metric of complexity. The rise from lower vertebrate to annelid or arthropod must have been accompanied by the evolution of differentiation of many cell types. This evolution would presumably be accompanied in turn and indeed accomplished in part through the establishment of the hierarchy of pattern-formation genes that mediate the development of metazoan architectures. Once such complexity was established, radiation of modified architectures might be relatively rapid. It is thus possible that the Vendian was a time of relatively gradual evolution of complexity, but that across the Precambrian-Cambrian transition a number of lineages reached independently that level of complexity permitting the establishment and rapid radiation of body architectures that we recognize as phyla, classes and orders (Valentine 1991). Jacobs (1990) has shown that early segment clades radiated more extensively than nonsegmented ones, and that mutations to homeobox-containing selector genes might provide just the sort of body plan variations that are common during the Cambrian, among segmented forms.

The problem of the origin of body plans is an area of rapid conceptual advance today, fueled by the applications of the techniques of molecular biology and by interpretations of the fossil record. Models, such as outlined above, would not be possible without input from both neontology and paleontology, and I foresee much more work in this area in the coming decade and beyond.

The Successive Dominance of Clades

From the rapidly radiating metazoans of the early Cambrian, the trilobites emerged to dominate the durably skeletonized fraction of the benthic faunas of the Cambrian Period. Thereafter the trilobites declined and were replaced by other groups as dominants in benthic communities. Sepkoski (1981) identified three major Phanerozoic marine invertebrate faunas: Fauna I, the trilobite-dominated Cambrian; Fauna II, in which crinoids and articulate brachiopods were most diverse and came to dominate the remaining Paleozoic; and Fauna III, in which gastropods and bivalves rose to dominate the Mesozoic and Cenozoic. Why did those clades, in that order, come to dominance?

There is an explanation available from fossil data, a hypothesis suggested by studies of extinction and speciation rates (see, for example, Stanley 1979; Sepkoski 1984; Van Valen 1985; Gilinsky and Bambach 1987). Empirically, different clades have different rates of speciation and extinction and therefore of taxonomic turnover (Simpson 1944; Stanley 1979; Holman 1989), and they maintain these differences over long periods of geologic time; the rates appear to be inherent in the biology of the clades. Trilobites, the dominant Fauna I taxon, had relatively high per taxon extinction rates (1.9 families/family/million years). Fauna II dominants, crinoids and brachiopods, had extinction rates of 1.3 and 1.0 families/family/million years, respectively. Dominant Fauna III taxa had still lower extinction rates (0.3 families/family/million years for both gastropods and bivalves). As all of these taxa are established during Cambrian time, it appears that the fast-turnover group was more successful in radiating to dominance early, but was then replaced by more extinction-resistant groups, and these in turn by still more extinction-resistant groups that diversified only slowly but that came to be the principal taxa in the long run. Table 2.2 displays the family extinction rates within the major groups during the times of dominance of each fauna in which their records are adequate. Extinction (and turnover) rates are declining through time *within* these principal

Table 2.2. Family extinction rates of major clades of Phanerozoic marine invertebrate faunas, per taxon per million years, during the times dominated by each fauna.

	Fauna I	Fauna II	Fauna III
Trilobites	2.8	1.6	—
Crinoids	—	2.1	0.8
Articulate Brachiopods	1.7	1.6	0.9
Gastropods	—	0.4	0.3
Bivalves	—	0.5	0.2

NOTES: Rates calculated from data of Sepkoski (1981) and supplements to February 1988 average extinction rate per state divided by 8 (the average stage duration in millions of years). (After Valentine et al. 1991a.)

taxa. Nevertheless, whenever dominant taxa of any of the faunas are found in association, their relative extinction rates follow the stated order. These data indicate that the clades with the most extinction resistance tend to become increasingly important through time.

There is a well-defined body of macroevolutionary theory that predicts just such an outcome. If extinction intensities are perfectly stable, a group with high speciation and extinction rates should survive just as well as a group with lower rates, but the fossil record clearly shows that extinction intensities have varied greatly over geological time. During episodic periods of high extinction, the extinction-prone groups would be more likely to be reduced to zero than the extinction-resistant groups, and the ability of the former to diversify rapidly would then be of no avail (see Raup 1978; Valentine 1989b, 1990b). Thus the general trend toward replacement of high-turnover by low-turnover clades is both expected and observed. These clade-characteristic rates are of course not adaptations per se but effects flowing from clade properties that were established probably during the early radiations that founded the clades. Such a property may have originated as an adaptation to meet some immediate requirement, and then as a by-product resulted in a modal level of resistance to change. At any rate, the composition of the living marine fauna would not seem to be largely a matter of chance, as implied for example by Gould (1989), although it probably does not result from superiority in ecological adaptations of the survivors. Rather, it results from coherent macroevolutionary processes.

The important question here is, What are the inherent clade factors that happen to confer successful turnover rates? It is doubtful that paleontologists will be able to figure this out without a lot of input from neontologists, who on the other hand could not even address the question without the paleontological evidence. My guess is that the key features are reproductive and developmental (larval) strategies, although the correlations are obviously not straightforward or there would be no problem. The answer could be of practical value—perhaps the data are, even as they stand—in conservation biology. When it comes to managing the marine biosphere it would clearly be useful to have some idea of which clades are historically the more vulnerable to extinction; it need not be those that are ecologically toughest. Also, although taxonomic triage is not a pretty concept, it may well become necessary to decide which species, among many belonging to different clades, we would most like to preserve. If such situations arise it will help to know what we're doing.

The Integration of Marine Communities

The third question that is expected to be of special future importance within evolutionary paleontology is rather ecological but has not yet received much formal study in marine ecosystems. The question is the widely discussed one of whether communities are very tightly integrated (by evolution) or are quite loose, flexible species associations, to contrast the more extreme options. Marine communities of the last million years of the Pleistocene are composed chiefly of species that are still living, and whose present distributions and associations are fairly well known. The gastropods and bivalves (dominant durably skeletonized clades at present) of the Californian Province are relatively well known, as is their Pleistocene record. Many of the Pleistocene fossil associations are on marine terraces, for which the depth of deposition, configuration of the shoreline, and general conditions of accumulation can be worked out and the fossil situation compared with that of the living fauna. The best known fossil faunas are from about 80,000 and 120,000 years before the present, although a number of other ages are represented also. The dates of the terrace assemblages are times of high sea-level stands, interglacial ages when glacial melting was near its peak for that time; they alternated with times of low sea stands when glaciers were extensive. But very little record of benthic communities is available for those times, as the deposits are chiefly below sea level, and much of the marine record that does exist is poorly dated.

In general it would be expected that the higher high sea stands would be the warmer interglacials, while the lower high sea stands would represent the cooler interglacials (both much warmer than glacial times, of course). The faunas of the interglacial ages of lowest ice volume (judging from isotopic evidence; for example, see Shackleton and Opdyke 1973) do indeed contain many warm-water species that live only south of their fossil location today, together with, surprisingly, some cool-water forms that live only to the north. The faunas of the interglacial ages with more ice volume have only cool-water forms. In both of these cases the bulk of the fauna is thermally ''normal'' for the area today. It appears that the fluctuating Pleistocene climates have driven the biogeographic changes. The number of mobile species is fairly large; of 649 living species known in the Californian Province during the last million years or so, 113, or about 17%, do not live today in the province at all but are found southward in the Panamanian Provincial region or northward in the Oregonian Provincial region. Many other species that end their ranges in the Californian Province today are found well outside of their present living ranges in Californian Pleistocene assemblages; I do not have a precise count, but a survey of some of the richer localities leads to an estimate that about 15% of the Pleistocene species belong to this element. Thus perhaps 32% of the Californian Pleistocene fauna has been demonstrably mobile. These data are chiefly from interglacial associations that experienced climates somewhat similar to today's. The glacial faunas, if well known, surely would reveal many other range shifts, for the climate then must have been quite different from that at present. There is no obvious reason that the mollusks should not be representative of the fauna as a whole; other taxa should have been equally mobile. Those other taxa that are partially known (bryozoans, decapods, echinoids, etc.) do display similar patterns of range shifts.

The point of citing these data here is that the communities that were subjected to so many arrivals and departures, that have undergone such extensive and repeated compositional changes, seem to have suffered no particular ill effects (Valentine and Jablonski, in prep.). The cumulative percentage of extinct species during the last million years is 16%, which plots right on the average Neogene curve of cumulative extinctions for the last twenty million years or so (see Stanley 1979), dating back far beyond the onset of the highly fluctuating climates. The mobile species seem to have been successful within their communities; species were accepted into or departed from trophic webs and managed new competitive interactions without any obvious ill effects. From reconstructions of the Pleistocene paleocommunities it seems that mobility was general and not restricted to any special environment, occurring from marginal marine marshes to outer shelf environments, and involving all trophic levels and habits that are well represented by mollusks (and the gastropods cover quite a spectrum). If there was any bias in these regards it is not clear from present evidence. The marine Pleistocene record of the tropics also displays faunal changes and shifts, but is not as well studied as yet. It is nevertheless clear that the Pleistocene changes are not restricted to habitats in highly variable conditions, but occur also in what are usually regarded as stable biotopes.

The fossil evidence seems to indicate that marine communities are quite robust with respect to some sorts of climatic and biotic change. This is not trivial information, and I believe that there is a promising area of future research here, in view of the expected climatic changes driven by human activities. The data cited above are "found" data in the sense that they were accumulated incidentally insofar as community theory is concerned. Dedicated studies should throw much additional light on this subject, and again, a mixture of paleontological and neontological data and expertise is most appropriate.

Conclusions

The three problems that I have addressed here are those with which I happen to have had some experience and can discuss with a less-than-average chance of making some major gaffe. These problems do seem ripe for important work in the coming decade, but they are by no means the only such fields that may blossom shortly; a large number of problems are characterized by requiring a combination of paleontological and neontological expertise. Let me recap the three problems very briefly. The Precambrian-Cambrian transition was a time when large numbers of body plans appeared. It seems likely that the body plan novelties were evolved for their own sakes, in the sense that they were not simply effects of some low-level process such as speciation (except trivially). There must have been a population genetics of the changes involved, but it would appear that we are dealing with pattern-formation genes and any clear understanding of what went on still lies ahead. Among the body plans of the Cambrian, a number displayed significant staying power and their lineages came to dominate marine communities. It can be empirically demonstrated that groups with low turnover rates came gradually to replace those with higher rates. The modern fauna owes its higher taxonomic composition to those processes. Nevertheless, we do not understand the biological bases of turnover rate regulation, although there are clues (see Stanley 1979 for suggestions). Opportunities abound. Finally, paleo-communities appear to be quite forgiving of species invasions or withdrawals, supporting the view that community structure is not fragile, and that communities are probably not "full" in the sense that there are no resources that are available or that can be released to support additional species. The circumstances and reasons for the invasability and the limits thereon are clearly of great concern both scientifically and practically.

Paleontology, it would seem, is merely a branch of biology. It has its own special tools and approaches, but in that regard it is like all branches that require special conceptual models and ancillary supporting sciences. In the case of paleontology, geology is an important ancillary science. Of course some paleontologists work entirely within the geological sciences, untangling stratigraphy and interpreting depositional frameworks, in which case paleontology is a servant of geology. When paleontology is practiced for itself, as a record of ancient life and ancient biological processes, the servant is on the other hand, and geology becomes a fundamental paleontological tool. I foresee a continued merging of paleontology with biology to produce a discipline that, in co-opting techniques of other branches of biological sciences, can attack and solve problems that are otherwise intractable. And I believe that paleontology is capable of returning the favor. Just as questions that are raised by fossil data must be answered chiefly through knowledge of processes investigated in the living world, so the explanations of the neontologist must pass the tests imposed by the fossil record of the history of life.

Summary

Three problems that will require a blend of paleontological and biological study for their solutions are outlined. One problem is the origin and early radiation of metazoan body plans. Fossil evidence suggests that bilaterans were represented by creeping worms until near the Cambrian-Precambrian boundary, when higher invertebrates arose explosively (if any higher invertebrate body plans arose earlier, they were most likely arthropodlike). Molecular evidence suggests that many bilaterian clades branched while at lower invertebrate grades. Some of the diverging lineages that produced higher invertebrate body plans arose independently. A second example involves taxonomic turnover rates among different phyla and classes. The taxa with the higher turnover rates became established first as dominants (Phanerozoic Fauna I) but gradually were replaced by more extinction-resistant taxa (Phanerozoic Fauna II) and finally by still more extinction-resistant forms (Phanerozoic Fauna III). The causes of the turnover-rate variations, which are empirical, are not understood. A third example involves the composition of marine benthic paleocommunities, which are altered more or less continuously and extensively during the Pleistocene without any increase in "background" extinction rates. These changes in community composition and, inevitably, structure suggest an openness and flexibility at odds with some hypotheses of community ecology. Paleontology, with its historical data, should be regarded as a field of biology and should become increasingly integrated with it in coming decades.

Acknowledgments

This paper draws from findings of NSF Grants EAR 84-17011, 90-15453, and 91-96068.

3

Five Properties of Environments

GRAHAM BELL

Ecology is the biological science of the environment; and after nearly a century of scientific ecology we know a great deal about a wide range of environments. What we still lack is any general theory of the environment. In genetics and evolutionary biology, the accumulation of knowledge is guided by a few simple and powerful theories. The equally vast accumulation of knowledge about environments seems to have proceeded in the virtual absence of general theory. Some areas of ecology, as it is usually understood, are indeed organized around clearly defined principles: demography is an obvious example, and some aspects of the study of material fluxes have now been handed down to the engineers. But I think that it is fair to say that our interpretation of the environment—surely the core of the discipline—remains undirected by any central organizing principles.

Organisms may respond to their environment by evolving sufficient phenotypic plasticity to cope with a wide range of circumstances. Alternatively, selection may instead favor specialization in the form of specific genetic adaptation to particular sites or conditions. In a uniform and unchanging environment, plasticity and diversity—to say nothing of dormancy, dispersal, and sexuality—would rarely evolve. To understand these basic properties of organisms, we must understand the environment to which they are a response. To understand them quantitatively, we need not only general but also quantitative theories of the environment.

In this essay, I shall offer a series of simple generalizations about environments which might be useful in formulating general theory. I shall begin with well-established facts, and conclude with untested speculations. Because my brief is to indicate directions for research in the next decade, most of what I have to say will refer to work in which I have been involved; it would be difficult to justify working along lines that you thought were unpromising. This chapter is in no sense a review of the topics that I shall discuss, and when I cite my own work it is not to claim either priority or superiority. For example, a similar argument is developed at much greater length by Tilman (1987), whose book should be consulted for a comprehensive review of the literature.

The five properties of environments that I shall discuss are as follows:

1. Environmental variance is relatively large.
2. The environment is complex on all scales in space and time.
3. The response of organisms to environment is indefinitely inconsistent.
4. Environmental variation is largely self-regulated.
5. Environments tend continually to deteriorate.

The conclusion this discussion will lead to is that environments are not the static or arbitrarily changing background to evolution, but instead are themselves dynamic and evolving entities.

1. Environmental Variance Is Relatively Large

By far the most extensive and most precise information that we possess about environments and the way in which organisms respond to them comes from agronomy. The distribution of variance among environmental and genotypic effects is shown in figure 3.1, which summarizes the results of a large number of agronomic trials covering a wide range of characters and crop species. It is clear that variation among environments, which are represented in such trials by different localities or years, greatly exceeds the variation contributed by genotypes (cultivars or experimental lines) or by genotype-environment interaction. If we narrow the focus by including only studies of seed yield, the character closest to Darwinian fitness, in which all three components of variation can be estimated, the same conclusion emerges (figure 3.2). Since every effort is made in trials of this sort to exaggerate the genotypic variance in order to facilitate selection and to minimize the environmental variance by soil treatment, fertilization, and pest control, the predominance of environmental effects is very striking.

Agronomic trials detect spatial variation on scales of about 10 km in manipulated environments. However, environmental variance is equally prominent at spatial scales down to 10 cm in undisturbed natural environments. Figure 3.3a is a map of soil nutrient levels in 50 m by 50 m area of undisturbed hardwood forest in southern Quebec, giving a vivid picture of the complexity of a natural environment on spatial scales relevant to the growth and dispersal of native plants. The varia-

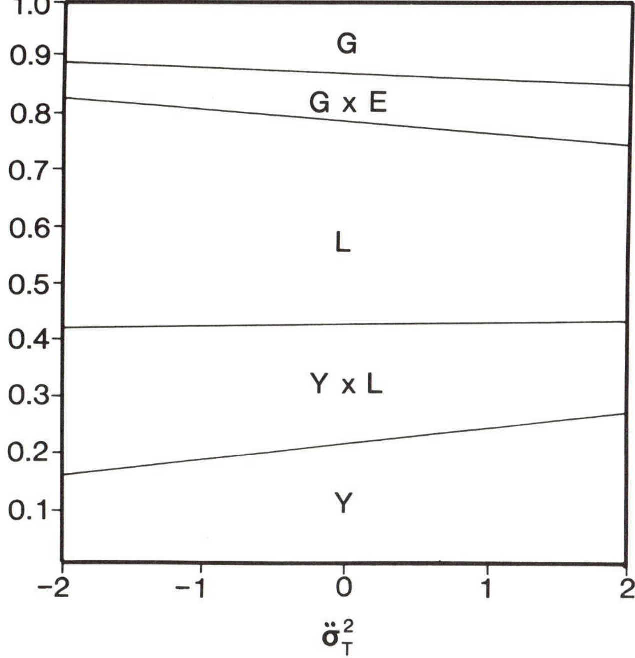

Figure 3.1. The overall composition of the phenotypic variance among crop plants for a wide range of characters. The ordinate is the total attributable variance, relative to the error variance. The vertical width of any of the five sections into which the diagram is partitioned then represents the average proportion of the total variance contributed by environment (year Y, locality L, year-locality combination Y × L), genotype (G), and genotype-environment interaction (G × E). (From Bell 1991a.)

tion that is detected by ion electrodes is expressed in plant growth. When seed families of the native herb *Impatiens pallida* are planted back into random sites within the area from which their parents were collected, the variance among sites greatly exceeds the variance among families (figure 3.4).

It is clear from these results that ecology is much more important than genetics: most of the variance that we see in natural populations, and in particular most of the variance of fitness, is associated with differences among environments rather than with differences among genotypes. The importance of genetics reappears when we consider the cause of this discrepancy. It is easy enough to contrive situations in the laboratory in which genetic variance predominates. However, when any such

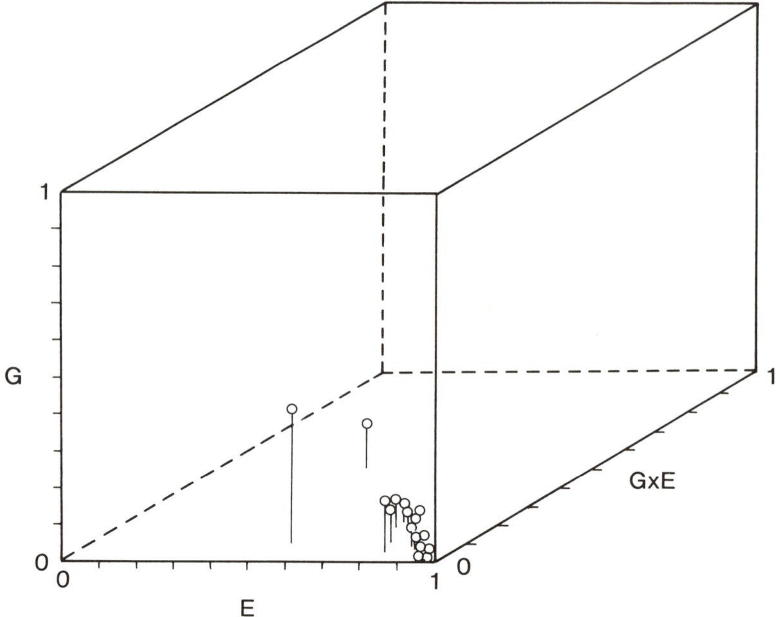

Figure 3.2. A three-dimensional plot of the genetic, environmental, and genotype-environment interaction variances of seed yield per plot in a number of agronomic trials. (From Bell 1991a.)

situation occurs, selection will act so as to consume genetic variance, either eliminating it entirely or converting it into other forms of genotypic variance, such as genotype-environment interaction. The prevalence of environmental effects both in managed and in natural systems can therefore be understood in very simple terms as the consequence of artificial or of natural selection.

2. *The Environment Is Complex on All Scales in Space and Time*

Despite the overall magnitude of environmental variance, one might hope that at some scale the environment is simple and uniform, licensing for ecology the reductionist approach that has been so successful in genetics. Unfortunately, this is not the case. Figure 3.3 maps soil potassium at scales of 50 m, 10 m, and 1 m in forest soil; at smaller scales the overall variance is less, but the environment seems to be just as complex. The simplest and perhaps the most useful way of expressing this result is to regress the variance of samples on their distance apart.

Similar techniques are routinely used by geostatisticians to express the increase in variance of physical variables with distance on a regional scale. We have measured the growth of genetically uniform plants to integrate those features of the environment which contribute to plant performance, and shown that biotic variance increases with distance at small spatial scales in undisturbed natural environments (figure 3.5a). When measurements are taken over many orders of magnitude, log-log plots of variance on distance seem to be linear, the variance increasing indefinitely with greater distance (figure 3.5b).

The increase in variance with distance implies that nearby sites are more similar than widely separated sites or, in other words, that the correlation among sites falls off with distance. We have found not only that this effect can be detected at small spatial scales, but also that the rate at which correlation falls off is independent of the absolute scale (figure 3.6). Natural environments thus have the chaotic, or fractal, property that complexity is retained at all spatial scales. No matter how powerful the magnification or how precise the measurement, simplicity and uniformity fail to appear.

Although these conclusions are based on a study of spatial variation, they may apply with equal force to temporal variation (see Pimm and Redfearn 1988), so that the environment becomes increasingly uncertain as one moves farther away either in space or in time. Figure 3.7 illustrates this tendency for a large-scale agronomic trial in which a single inbred line of barley was sown into a single field divided into a large number of replicate plots over a series of years, and the seed yield of each plot was recorded. The figure expresses the results of the trial in the form of a spatiotemporal autocorrelagram, which shows how the correlations of seed yield among plots falls off in space and time. It can be thought of as a map of the environment that will indicate the likely fate of propagules that are being dispersed in space and time, and which will not only influence the dynamics of the population as a whole but will also effect the evolution of plasticity, the maintenance of genetic diversity, and the evolution of characters such as dispersal, dormancy, and sexual diversification.

3. The Response of Organisms to Environment Is Indefinitely Inconsistent

The genetic structure of lineages has the same property of indefinite complexity as the environments they inhabit. If any twig of the phylogenetic tree is magnified, the overall phenotypic variance decreases as

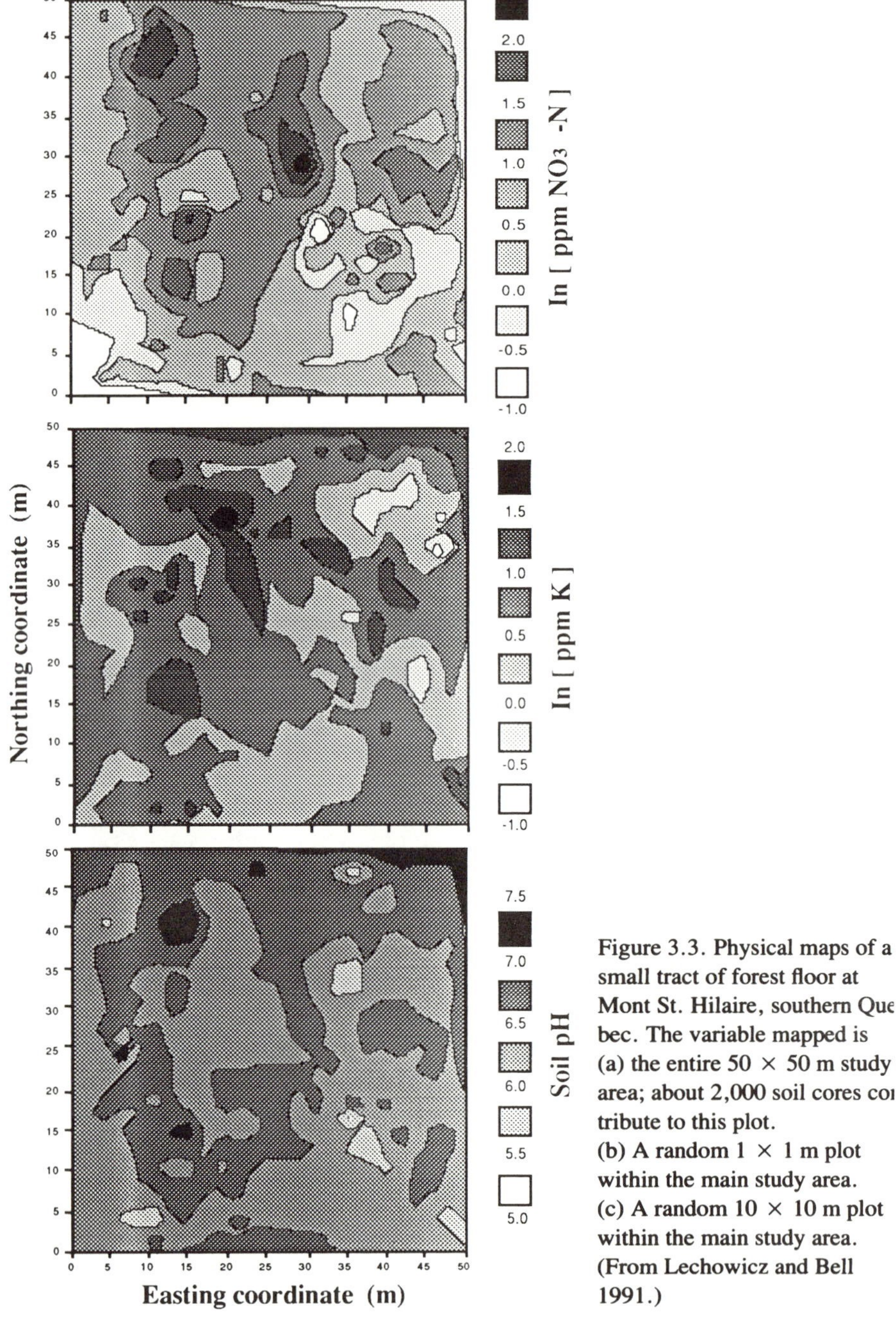

Figure 3.3. Physical maps of a small tract of forest floor at Mont St. Hilaire, southern Quebec. The variable mapped is (a) the entire 50 × 50 m study area; about 2,000 soil cores contribute to this plot.
(b) A random 1 × 1 m plot within the main study area.
(c) A random 10 × 10 m plot within the main study area.
(From Lechowicz and Bell 1991.)

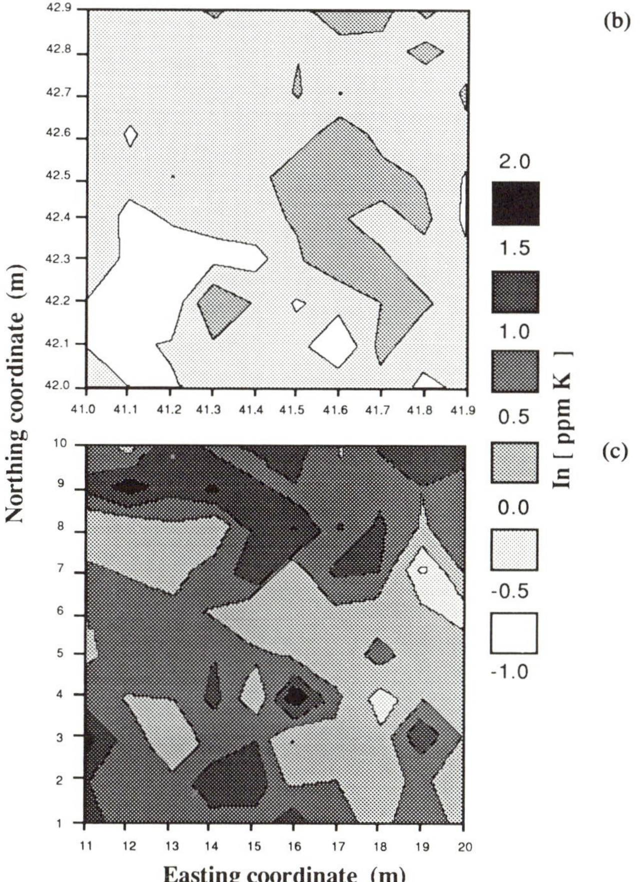

one shifts from phyla to genetic families, but the complexity of the branching structure remains unchanged. In natural systems this complexity is expressed against the complex environmental background I have just described. The outcome will depend on whether genotypes (at any scale, from phyla to clones) respond consistently to environments (at any scale, from regions to the scale of the organisms themselves). If organisms respond consistently to environmental variation, then one type of organism will generally be superior; genetics and ecology are effectively decoupled because genetic or phylogenetic diversity cannot be explained through environmental diversity. If organisms generally respond inconsistently, so that the ranking of performance of different types varies widely among environments, then genetics and ecology are effectively the same subject, whose separation can be defended only in

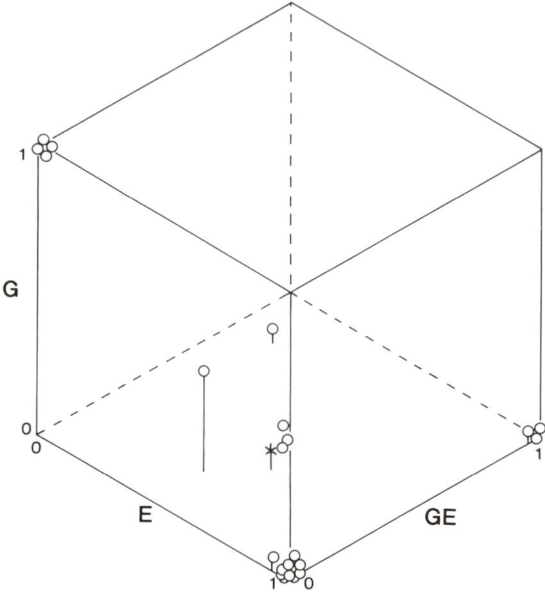

Figure 3.4. Environmental and genotypic sources of variance in fitness in a native population of *Impatiens pallida*. Each point represents a set of crosses among four random parents from the 50 m by 50 m area shown in figure 3.3, planted back into four random sites within this area. "Fitness" is total fruit and flower production. The asterisk is the mean over all twenty-four sets of crosses. (From Bell, Lechowicz and Schoen 1991.)

special cases as a matter of convenience. I shall argue that the latter point of view is closer to the truth than the former: just as there is no scale at which the environment becomes uniform and unstructured, so too there is no scale at which response to environment becomes consistent.

This issue can be settled most readily by studying the population growth of small organisms whose environment and genotype are easily manipulated. I have used the unicellular green alga *Chlamydomonas*, which displays rapid and repeatable logistic growth in batch culture. The results of some simple factorial trials are shown in figure 3.8, in the form of norms of reaction, which are regressions of the performance of a genotype on a measure of environment: if the norms of reaction are parallel, the response to environment is consistent. Instead, the norms of reaction frequently cross one another, demonstrating that the response to environment is inconsistent, whether genotypes involved are distinct species or full sibs from the same family.

We can extend this result by asking how the extent of inconsistency depends on the variation among environments. If two environments are very similar then the ranking of genotypes should also be very similar, so that the genetic correlation between performance in one environment and performance in the other is close to unity. It is natural to expect that as the environments become more different, the genetic correlation will fall. In the *Chlamydomonas* trials, the environments are constructed as factorial combinations of levels of three macronutrients, and may differ with respect to the levels of one, two, or three macronutrients. The genetic correlation is lower when the number of factors varied is greater (figure 3.9). When many environments are constructed, it is possible to show that the genetic correlation declines continuously with environmental variation, approaching zero for environments that are sufficiently different (figure 3.10). Crop plants show the same behavior in agronomic trials (figure 3.11).

To demonstrate the extreme complexity of environments and of the response of organisms to their environment is not a confession of failure nor a justification of obscuranticism. It is true that simple reductionist techniques and models are not likely to be very successful, but it should not be surprising to realize that statistical principles are not merely convenient devices for analyzing data, but also represent the processes in the natural world which generated the data. In the sections above I have tried to establish two statistical generalizations that connect the properties of environments with the properties of organisms: first, that environmental variance increases log-linearly with distance in space and time; and secondly, that genetic correlation decreases toward zero as environmental variance increases.

4. Environmental Variation Is Largely Self-regulated

So far, I have discussed environments as though they were fixed constraints to which organisms respond. In most cases this is fallacious and is adopted only for convenience: the argument becomes much more subtle when we recognize that organisms can themselves act as environmental factors. Physical and biotic factors are fundamentally different. Organisms can readily adapt to either, but biotic factors can themselves adapt. When the environment is Darwinian, evolution becomes a mat-

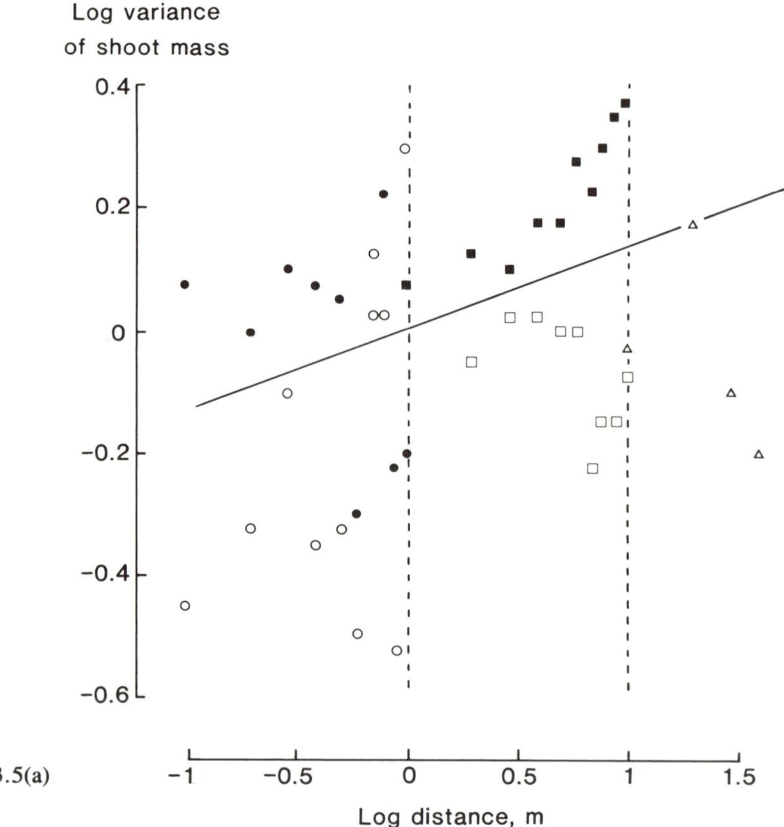

Figure 3.5. The increase of environmental variance with distance. (a) The measure of environment is the total dry weight of standard cultivars of barley planted into soil cores taken from the same forest site shown in figures 3.3 and 3.4. Different symbols refer to different replicate plots at a given scale. (From Bell and Lechowicz 1991). (b) Surveys of the spatial structure of soil pH over eight orders of magnitude, from 10 cm to 10,000 km. The hollow triangles represent data from the 50 m by 50 m area at Mont St. Hilaire referred to in the text. (For full documentation, see Bell et al. 1991.)

ter, not merely of improving adaptation, but of clinging on to current levels of adaptedness.

A factorial experiment in which all pairwise combinations of sibs obtained by crossing isolates of *Chlamydomonas reinhardtii* are cultured enables us to calculate the average effect of a genotype on the production of mixtures in which it is a component. The average effect in mixture is highly correlated with performance in pure culture (figure 3.12a). We infer that interactions among genotypes are weak or non-

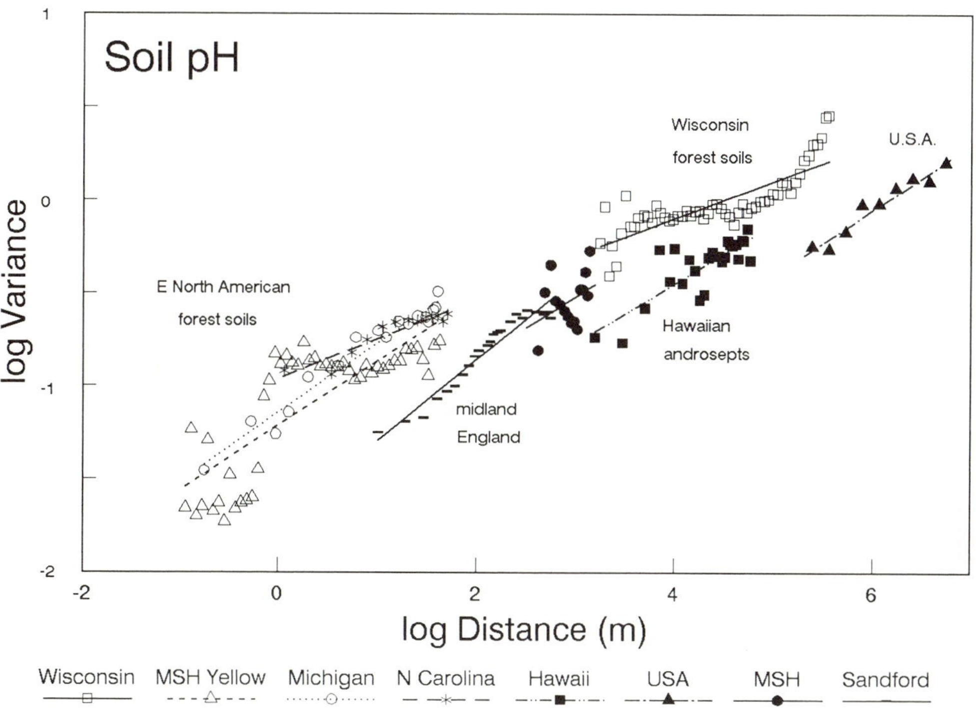

3.5(b)

existent, and in fact it can be confirmed experimentally that in most cases the genotype which is the higher-yielding in pure culture simply replaces the lower-yielding genotype during the growth of culture. When different species are cultured together, the result is quite different. The average effect in mixture is uncorrelated with performance in pure culture (figure 3.12b), suggesting that there can be powerful interactions among genotypes. This can be confirmed experimentally by growing genotypes in a medium that has been conditioned by the previous culture of the same or of another genotype. Such experiments show that a given genotype may have strong positive or negative effects on the growth of other genotypes, so that mixtures may consistently exceed or consistently fall below the mean performance of their components in pure culture. Thus, the presence of other types of organism may constitute an important component of the environment, provided that they are sufficiently different from the target organism.

The importance of competition has been a commonplace of ecological theory for so long that I will not belabor the point. It implies that

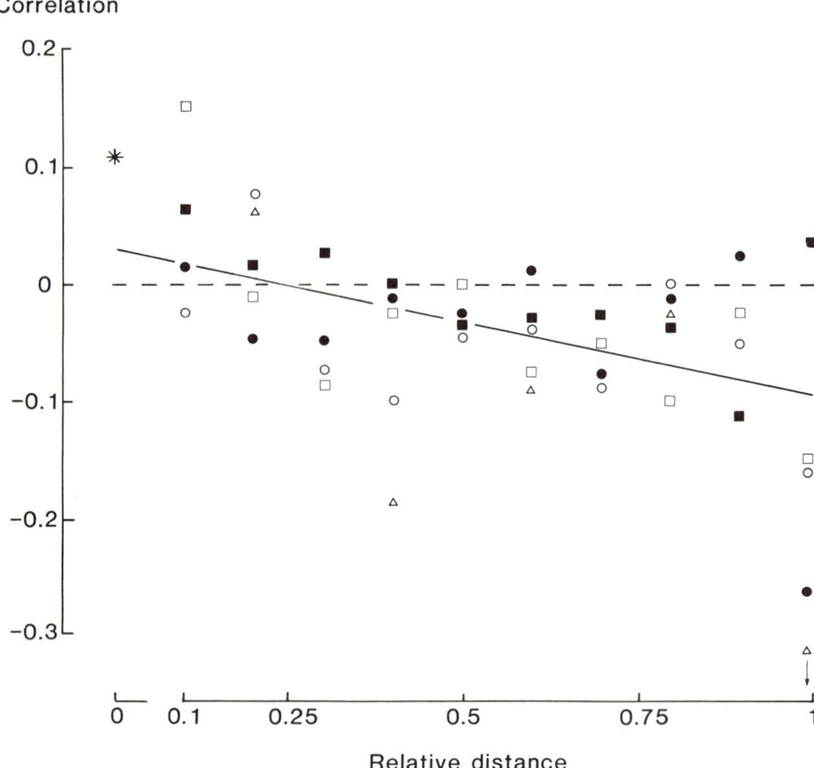

Figure. 3.6. The decrease of correlation with distance. This refers to the same data as figure 3.5. Note that the distance scale is relative, representing one-tenth of the overall grid dimensions of 50 m (triangles), 10 m (solid circles), and 1 m (open circles), respectively. (From Bell and Lechowicz 1991.)

environmental heterogeneity is due in large part to the diversity of other organisms with which the target organism interacts, and indeed, although no experimental demonstration seems to exist, it is difficult to imagine how the variance of natural environments on small spatial scales could be generated or sustained in any other way. A slightly more original point can be made by noting that, in the last experiment described above, the environmental effect of an organism, in conditioning the medium in which it grew, persisted through time, so that the growth of a genotype was affected by the previous occupant of the culture. Indeed, it was found that strong effects could be extended through three cycles of conditioning. A similar phenomenon is, of course, well known in agronomy, where it has given rise to the practice of crop ro-

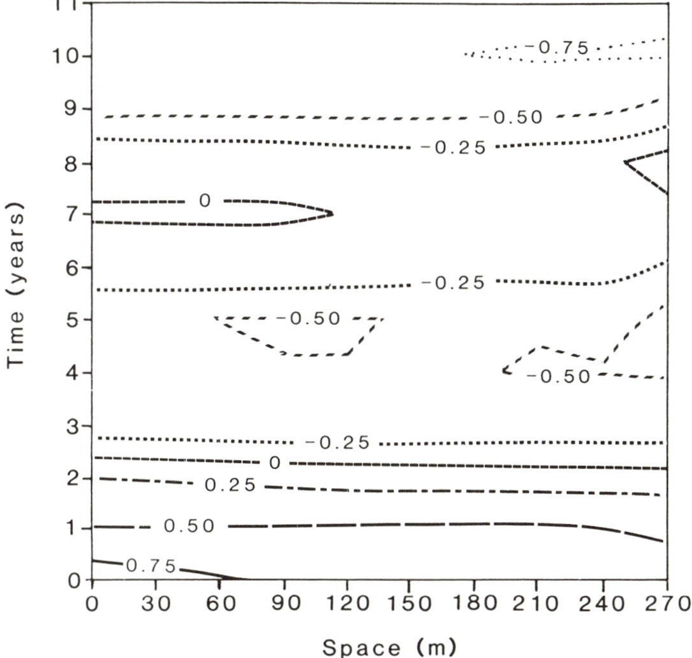

Figure 3.7. Spatiotemporal autocorrelegram for a barley uniformity trial. The response variable is grain yield per plot. Contours represent lines of equal correlation for plots separated by a given distance in time and space. (Data from Baker et al. 1951. From Bell 1991a.)

tation. Such effects seem bound to occur when resources are renewable, and when they can be depleted faster than they are renewed. A type that requires large quantities of a resource will reduce the availability of that resource in the future, and thereby enhance the future fitness of an alternative type which requires lesser quantities of the same resource. It is well known that contemporary interactions may lead to the persistence of many genotypes within a sexual species, or of many species within a community, through the complementary usage of resources which are not depleted. When resources are renewable, which is to say depletable, such interactions are lagged in time, the effect of which is a tendency to destabilize equilibria. Rather than having a stable composition, populations or communities may fluctuate in composition in a cyclical or chaotic fashion through time.

Thus, interactions among organisms may lead to a process of continual change. The environment may be dynamic, not because of forcing by external physical factors (although this certainly occurs), but because

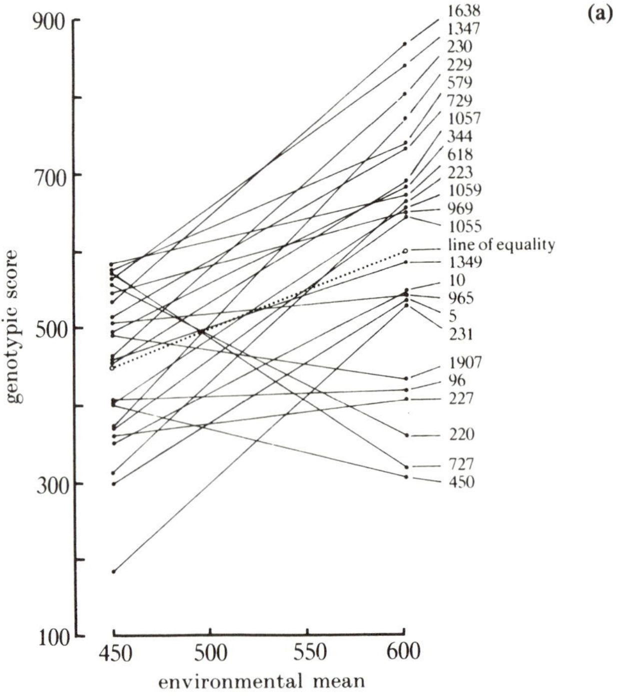

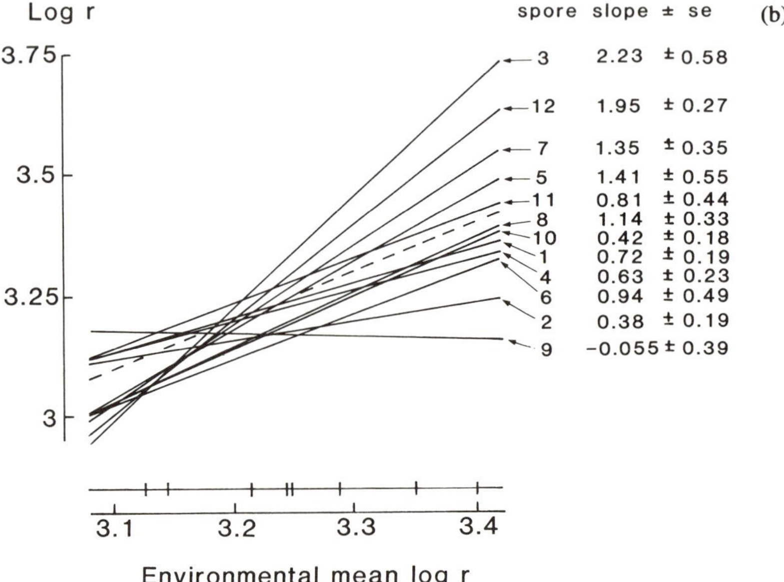

Figure 3.8. Norms of reaction in *Chlamydomonas* cross extensively. The response variable is the carrying capacity K of the logistic model. (a) Different species (from Bell 1990a). (b) Spores from crosses within *C. reinhardtii* (from Bell 1991b).

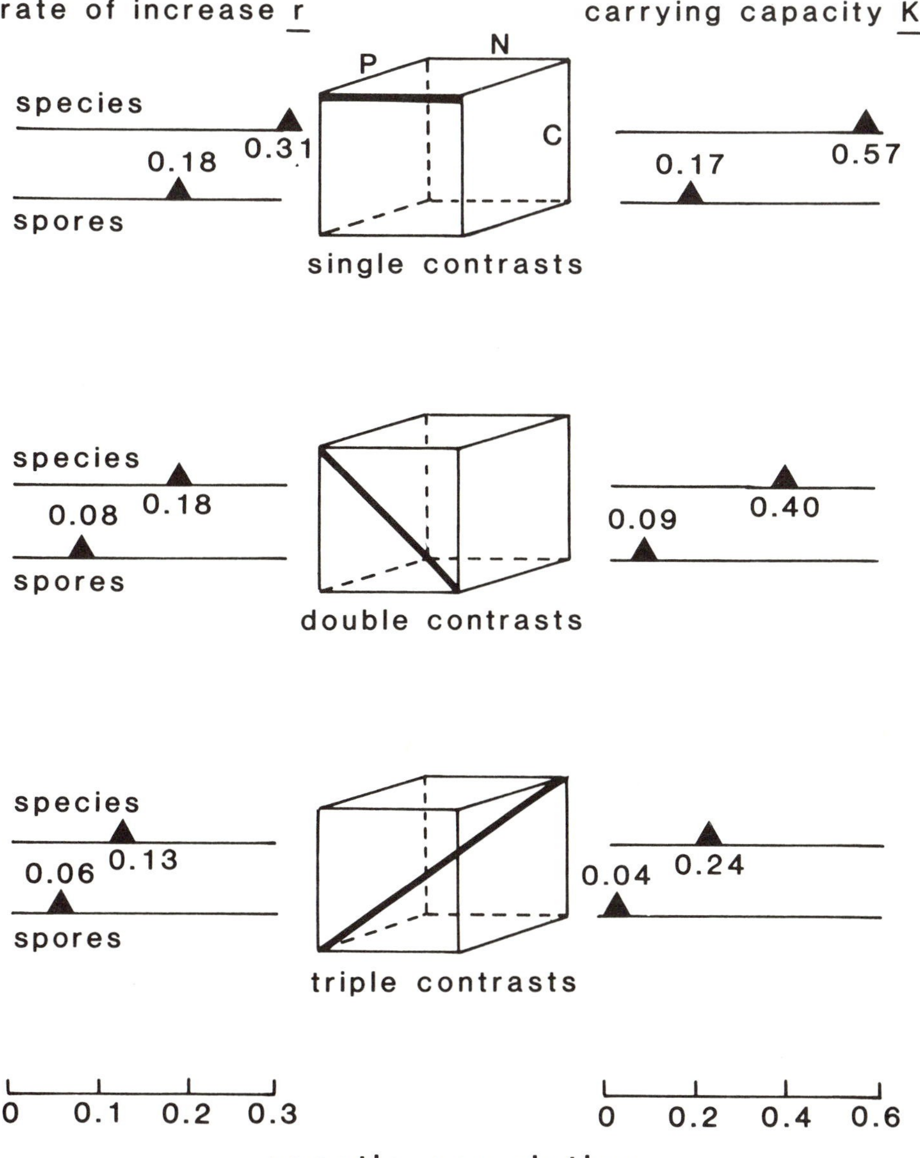

Figure 3.9. As more physical factors are varied, the response of genotypes becomes less consistent. Physical factors are levels of nitrate, phosphate, and bicarbonate in a factorial experiment with *Chlamydomonas*. The measure of consistency is the (pairwise) genetic correlation between environments, either for species or for spores from crosses within *C. reinhardtii*. (Data from Bell 1990a, 1991a.)

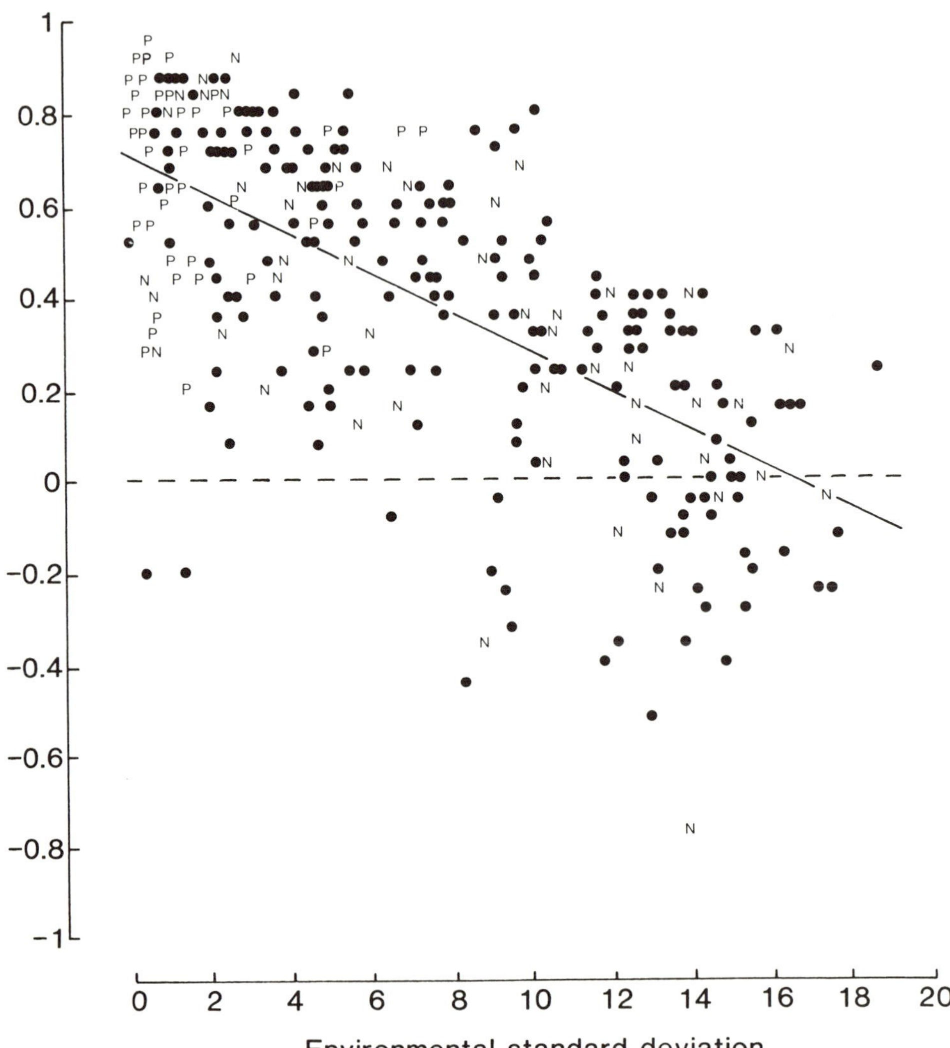

Figure 3.10. The genetic correlation between environments decreases with the difference between the environments. Each point represents a pair of environments for which the genetic correlation among fifteen genotypes (species) of *Chlamydomonas* and the variance of the two means has been calculated. The response variable is production after ten days of growth. The environments are factorial combinations of five levels each of nitrate and phosphate, giving ½(25 × 24) = 300 pairwise combinations of environments. (From Bell 1991d.)

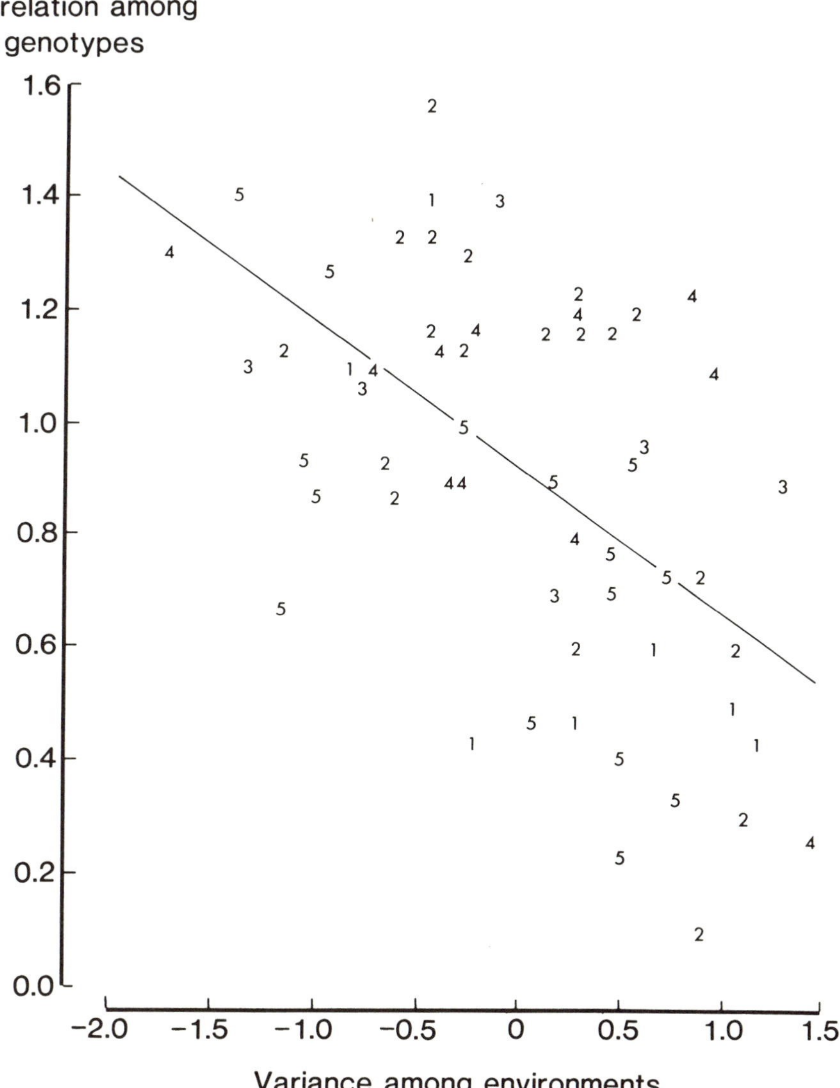

Figure 3.11. The genetic correlation between environments decreases with the difference between the environments. Each point represents an agronomic trial conducted over a range of environments for which the intraclass genetic correlation coefficient (genetic variance divided by the sum of the genetic and genotype-environment interaction variance) and the variance of environmental means has been calculated. The numbers plotted refer to different categories of characters: 1, seed yield; 2, components of seed yield; 3, vegetative yield; 4, morphology; 5, chemical composition. (From Bell 1991a.)

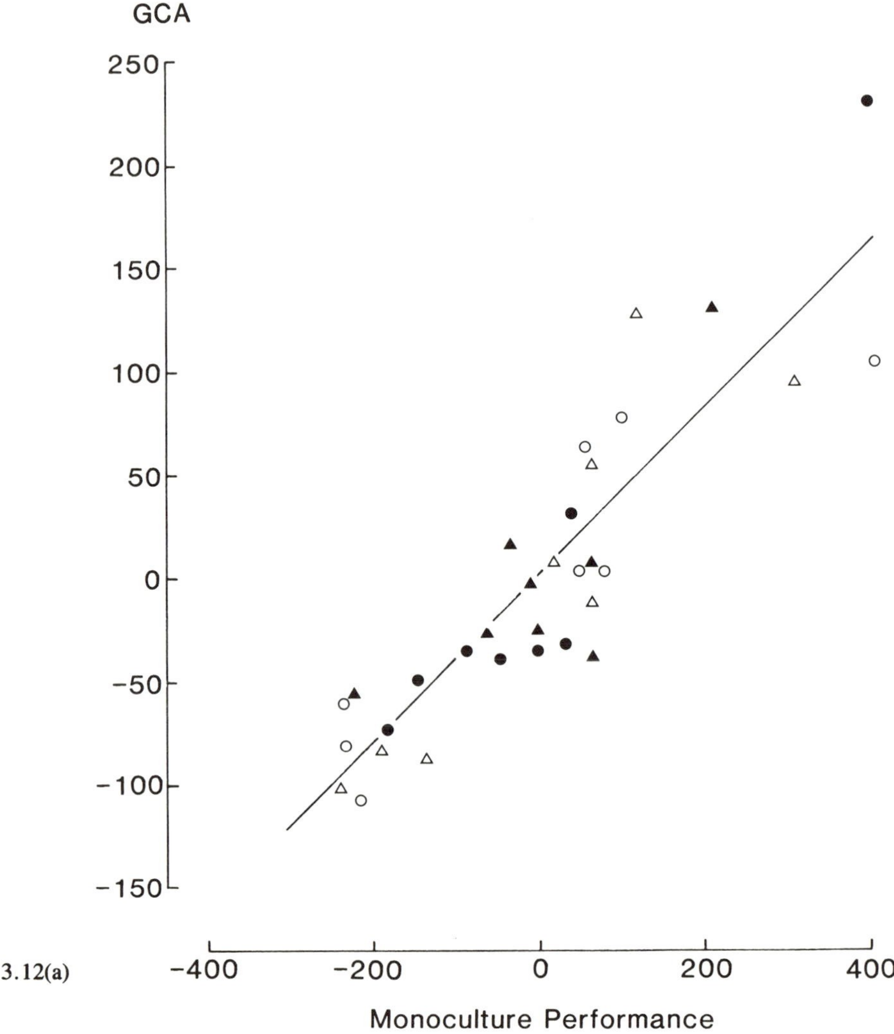

Figure 3.12. The relationship between performance in pure culture and performance in mixture. (a) Performance in pure culture is highly correlated with performance in pairwise mixtures of *Chlamydomonas reinhardtii* half-sibs (from Bell 1991c). (b) Performance in pure culture is uncorrelated with performance in pairwise mixtures of *Chlamydomonas* species (from Bell 1990b).

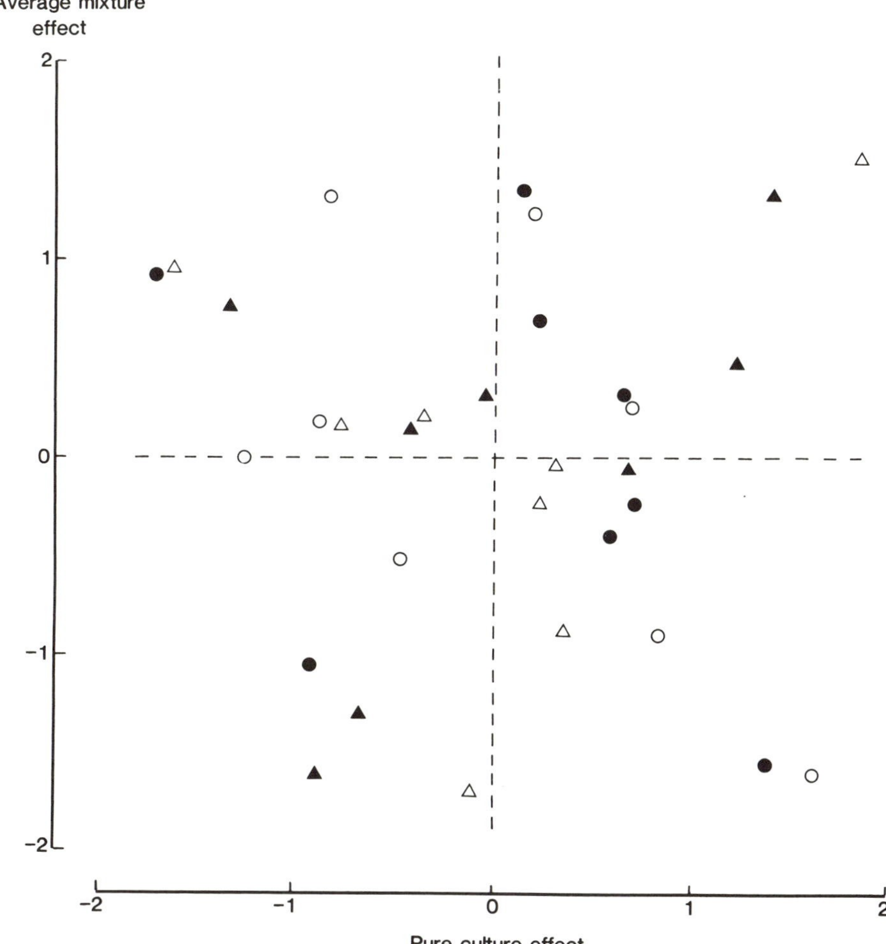

3.12(b)

of the properties of organisms themselves. Once we admit that organisms may themselves act as environmental factors, we seem to be led to conclude both that organisms themselves construct environmental heterogeneity and that they perpetuate a continually shifting pattern of heterogeneity. This is an unproven but probably not very controversial speculation. It can be taken a step farther by making a stronger and more vulnerable claim: that the environment is not only continually changing, but also that this change has a preferred direction. Environments tend to get worse.

5. Environments Tend Continually to Deteriorate

The most likely source of intense and specific interactions that are lagged in time is not the differential utilization of renewable resources by similar organisms, but directly antagonistic relationships in which the organisms themselves constitute resources for one another, and especially the relationships between pathogens or parasites and their hosts. The time lag arises from the delayed response of species of pathogens to changes in the abundance of their hosts (and vice versa), or through the delayed response of genotype frequencies within a pathogen population to shifts in the genetic composition of the host population (and vice versa). At first sight, it is not obvious that the bad news should predominate; after all, mutualistic interactions are perhaps as frequent, as intense, and as specific as antagonisms, and a more optimistic view of the world might be that it is continually improving. However, I do not think that this is likely to be the case. Selection will favor mutualists that cooperate more effectively with common types in their partner. The relative fitness of these common types will therefore remain unchanged; rare types will remain at a disadvantage. Among antagonists, the reverse is true. Selection will favor pathogens that can exploit common types in the host population more effectively, or hosts that can evade common types in the pathogen population more successfully. Consequently, the relative fitness of common types will always tend to fall through time, as a consequence of their being common, while the relative fitness of rare types will increase, until they in turn become common. The average individual is likely to belong to a common type, and will therefore perceive the environment as continually deteriorating.

There is not direct evidence, so far as I know, which would test this speculation. There is one major piece of indirect evidence: the prevalence of sex and genetic recombination in large and long-lived organisms. One of the few reasonable explanations of the systematic destruction of successful genomes is that the fruits of their success ensure their future decline. About a decade ago, a group of biologists inspired by Williams (1975) and Maynard Smith (1978) realized that time-lagged frequency-dependent selection created by antagonistic interactions would provide an environment in which sex made sense (Jaenike 1978; Hamilton 1980; Hutson and Law 1981; Bell 1982). It has since been established that this argument holds good in fully coevolutionary models (Bell and Maynard Smith 1987), and it seems likely that the argument will only be strengthened by increased complexity (Seger

1988; Hamilton et al. 1990). Whether or not the argument is true is undecided. Longer-lived species seem to have higher rates of recombination, as expected (Burt and Bell 1987), but the crucial comparative tests (whether species that are more highly parasitized have more recombination) has not yet been reported. There is some evidence that diverse populations are more resistant to pathogens (Kelley, Antonovics, and Schmitt 1988; Barrett 1981), but the crucial experimental test (that recombination increases as a correlated response to selection for pathogen resistance) likewise remains to be done. We are left with the intriguing possibility that one of the most important clues to the nature of environments is provided by the mating behavior of their inhabitants.

If I were to extract a single general message from the train of thought that had led to this conclusion, I would extend what is perhaps the single most profound analogy in ecology—the ecological theater and the evolutionary play (Hutchinson 1965). The play is enacted on a stage, and at any moment the design of the stage constrains the action within certain limits. Perhaps some of these limits cannot be transgressed: Agincourt cannot be crammed convincingly within a wooden O, and even a Roman amphitheater was not really capable of staging a convincing battle scene, let alone a voyage. Such limits are set by irreducible physical laws. Within these limits, however, what we see in the theater is the outcome of a continual interplay between the constraints imposed on the action of the play by the design of the stage, and the demand for changes in stage design created by the development of novel kinds of plays. There is a tendency to regard ecology as setting the limits within which selection operates, and regarding the rate and direction of evolution as being determined by wholly extrinsic factors such as marine recession or glaciation. There is, of course, a good deal of truth in this view. It is misleading only if it blinds us to the fact that the environment is to a large extent the product of living organisms, and therefore to an equally large extent the by-product of their evolution. Ecology and evolution are thus related, not as pattern and process, but rather as coupled processes of change, with every performance of the play redesigning the stage. The endlessly complex, shifting, and deteriorating nature of the environment arises from the activities of its inhabitants themselves.

Future Directions

These generalizations will not impress anyone who believes that the only valid function of theory is to provide designs for experiments or protocols for observations. To meet this criticism, I feel bound to suggest questions and problems arising from the argument laid out above that I think will be worth empirical investigation in the next decade. Specifically, I propose that it should be possible to make rapid progress in measuring the magnitude and scaling of environmental variance, and in conducting experiments on the direction of environmental change.

My first proposal concerns the description of environments. In the past, it has been usual to categorize supposedly distinct types of environment as habitats, to refer to environments as being coarse-grained or fine-grained, and so forth. This typological approach is now being replaced by estimates of environmental means and variances. The phenotypic variation of most organisms (large organisms, at least) is not smooth and continuous, but rather aggregated into the lumps that we call species. Are environments also lumpy, the lumps being what we call habitats? Or is environmental variation instead rather smoothly and continuously distributed, so that the concept of habitat should be discarded entirely? The mean values of environmental variables certainly change from place to place; does environmental variance itself vary from place to place? If so, what does the scatter plot of environmental variance on environmental mean look like? I have shown that, in one particular place, the environmental variance increases with distance; how does the slope of this fundamental scaling relationship change from place to place, and how does it depend on the particular variable measured? Environmental variance is relevant to organisms only to the extent that the variance that we detect from instrument readings is also manifest as phenotypic variance. Environmental variance can therefore be detected in bioassays only to the extent that organisms are not indefinitely plastic. To what extent are the phenotypes of organisms growing in their native environment less variable than instrument readings? If there is a cost of plasticity that prevents organisms from becoming perfectly plastic, what sets the limits to plasticity, and how does performance respond to selection for increased plasticity? Once we have quantitative descriptions of environmental heterogeneity, how is biotic diversity related to environmental variance? To describe this relationship more precisely, is it possible to express biotic diversity itself as a variance, perhaps through the frequency of different substances in some ubiquitous class of chemical compounds? If we deliberately manipulate the physical variance of environments, do more heterogeneous environments support more biotic diversity, or does selection instead favor greater phenotypic plasticity? What are the answers to all these questions if we investigate the temporal rather than the spatial structure of environments?

My second proposal concerns the interpretation of environments as evolving entities that, from the point of view of some target population, tend continually to deteriorate. It is technically feasible in closed experimental systems to expose a clonal target population, which can evolve only slowly if at all, to an environment inoculated with a large and variable population of antagonistic microbes. Such a target population would presumably perform less well than a control population maintained under sterile conditions; but would its performance continue to deteriorate relative to the control? If so, is the effect specific to particular target genotypes? That is, if target clones were reciprocally transferred to one another's environment, to what extent would the deterioration that had occurred be mitigated? A similar experiment could be attempted in natural populations. Suppose that a large number of individuals, replicated clonally under sterile conditions in the laboratory, were transplanted back into the natural environment from which their founder was isolated. After a period of time, these individuals are removed, and a fresh set is put out. In this way, the local population of antagonists would be presented with a large stationary target. Would the performance of the clone in the natural environment continually deteriorate relative to its performance under sterile conditions in the laboratory? If a genotype is stored for many generations, or if a natural clonal store can be found, is there a tendency for its performance, when grown in its native environment, to increase as a function of its age, relative to material grown in the laboratory?

This is naturally a partial and personal view of the direction in which ecology may progress. It will be worthwhile if it contributes to the emergence of a Darwinian interpretation of the environment: the evolutionary theater and the ecological play.

Summary

Ecology lacks any useful general theory of the environment. I put forward five generalizations as contributions toward a general theory, supporting them with observations in undisturbed, natural forest environments and with experimental results using cultures of *Chlamydomonas* in the laboratory. These five generalizations are as follows.

1. Environmental variance is relatively large. Environmental variance of fitness and related characters greatly exceeds genetic variance.
2. The environment is complex on all scales in space and time. Environmental variance increases indefinitely and without limit as distance increases.
3. The response of organisms to environment is indefinitely inconsistent. Genotype-environment interaction is all-pervading; the genetic correlation between environments falls as the variance of the environments increases.
4. Environmental variation is largely self-generated. Organisms are themselves important parts of each other's environment, and environments therefore respond and evolve.
5. Environments tend continually to deteriorate. Because antagonists tend to adapt to the common host type, the environment is consistently and continuously becoming worse from the point of view of most individuals.

I suggest two major research programs for the next decade: the quantification of environmental variation, and the investigation of environments as evolving antagonists.

Acknowledgments

I have distilled this chapter from collaborations with two colleagues: work on the scaling of environmental variance with Martin Lechowicz, and conversations on the nature of environmental change with Austin Burt.

4

Behavior and Evolution

MARY JANE WEST-EBERHARD

John Bonner (e.g., 1980, 1988) has often treated development and behavior together when discussing evolution. I used to credit this to the intrinsic fascination of behavior (my favorite subject), even for a developmental biologist. But the connection is clearly deeper than that, as Bonner's books show: behavior and development have a special common relation to selection and evolution.

A fundamental quality shared by behavior and development is condition-sensitivity: both are partly directed by circumstances. It is the curse of environmental influence that sometimes has made both of these topics seem awkward for evolutionists. Condition-sensitive processes are in some sense ''nongenetic''; and to be evolutionary is to be genetic: most modern evolutionary biologists define organic evolution as involving a change in population genotype frequencies. This raises the problem I will discuss: How, precisely, do condition-sensitive processes fit into a genetic theory of evolution? How do they evolve, and how do they affect evolution?

Behavior and Development: Divergence without Speciation

There is increasing attention to development in evolutionary biology, but I do not think that epigenetic considerations have really ''sunk in'' to the way biologists think. I believe, in fact, that we have a sort of ''collective blind spot'' against thinking effectively about flexibility. This blind spot is reflected in the way we discuss divergence: we learn—and teach—that character divergence requires reproductive isolation, or speciation. If you look in the literature you can find a whole forest of phylogenetic trees. They may be shrublike, cactuslike, twiggy, or gnarled. But each one associates divergence (or distinguishability) with speciation, or phylogenetic branching. If you consider what this implies in terms of the structure of populations, it has beneath it the assumption that populations are unimodally adapted. So, to get two adaptive peaks, or modes, you must have a branch.

To say the same thing another way, frequently heard: each species sits on a single adaptive peak. To get to another adaptive peak like that of another species, a population may have to pass through a valley of low fitness, which means a period of being poorly adapted and in danger of extinction.

A belief in unimodal adaptation is understandable because most characters measured, for example by taxonomists and paleontologists, show unimodal continuous variation. But the expectation of unimodal variation is self-fulfilling: a phenotypic discontinuity is taken to indicate a species or subspecies (reproductive) boundary, so that actual frequency of polymorphism is undoubtedly underestimated. The old typology that saw a species as a single morphological form has given way to a more sophisticated populational typology that sees it as a unimodal distribution of forms. This problem must be particularly acute in paleontology, where the kinds of field and laboratory observations that might reveal polymodal adaptation in morphology and behavior cannot usually be done. So it would not be surprising if paleontologists were to prove particularly susceptible to a perception of populations as unimodally adapted, and to the related idea that divergence requires speciation. Indeed, the idea that divergence requires branching is carried to an extreme by the paleontological hypothesis of punctuated equilibrium (Eldredge and Gould 1972): not only does it depict divergence as requiring speciation, but it holds that virtually all evolution occurs at the time of speciation, with stasis in the interval between.

Punctuated equilibrium (and the general association of divergence and branching) is consistent with a whole set of ideas that highlight the difficulty of change without speciation. Although each of these ideas was proposed independently to explain particular observations, taken together they constitute a many-faceted, coherent argument for stasis in the absence of speciation: a trait that is maintained by selection is not simply unchanging but may be subject to "stabilizing selection" (Schmalhausen 1949) and "canalization" (Waddington 1975) rendering it more resistant to change; change is further obstructed by the coadapted cohesiveness of gene complexes (Mayr 1963); populations are protected from disruptive gene flow (hybridization) by "reproductive isolating mechanisms" (Dobzhansky 1937, 1970; Mayr 1963); and evolutionary novelty requires a "genetic revolution" to overcome the adaptive genetic cohesiveness of a gene pool (Mayr 1963). Even the occurrence of behavioral options has been described in terms of stability and equilibrium, as in the terms "evolutionarily stable strategy," and "frequency dependent equilibrium" (e.g., Maynard Smith 1982). Dobzhansky (1970, p. 38) discussed divergent alternative traits (polyphenisms) and other kinds of phenotypic flexibility as examples of "developmental homeostasis," thus emphasizing the stabilizing rather than the diversifying effects of plasticity. Similarly, the evolution-directing effect of development is frequently discussed by present-day evolutionists in terms of "constraints" (Maynard Smith et al. 1985) rather than as "opportunities" for production of altered phenotypes.

In effect, we have constructed a theory of stasis, even though its proponents of course recognize that directional and diversifying selection may sometimes produce change; that behavior often takes the lead in evolution (e.g., see Mayr 1970, 1974); and that major steps in evolution often involve regulatory change (Gould 1977; Maynard Smith et al. 1985).

There is evidence to support all of these concepts, and I am not prepared to argue that the phylogenetic branching theory of divergence is wrong, only that it is incomplete. Geographic speciation, with spatial isolation promoting divergence and reproductive isolation (Mayr 1963) is among the best-supported hypotheses in evolutionary biology. Nor should this description of a "theory of stasis" be taken as a summary of all of evolutionary biology, or of "neo-Darwinism," or even of what any one person might say. Homeostasis is one result of plasticity; another is a capacity for change (see Thoday 1955 and the discussion by Dobzhansky 1970, p. 326). But the more frequent emphasis on the conservative result—stasis, stability, homeostasis, equilibrium, constraint—reveals a widespread conviction that evolution (and especially the origin of divergent novelties) may be difficult without speciation or the branching of lineages.

Here I will pursue the complementary idea that diversity can multiply not only by speciation, but also via developmental branching, especially in the form of intraspecific alternative phenotypes. Alternative phenotypes are common (for many examples, see references in West-Eberhard 1986, 1989). They exist side by side within populations; and they do not require reproductive isolation for their divergent evolution, only independent expression. This kind of divergence (developmental rather than phylogenetic branching) may lead to the fixation of one or the other alternative form in one segment of a population, giving the impression that phylogenetic branching has occurred when in fact the novel trait has originated as an alternative.

This idea is not new (see references in West-Eberhard 1986). It has been repeatedly suggested, and repeatedly dismissed or forgotten. I believe it gets dismissed because it requires a fundamentally different way of thinking about selection that touches and challenges the assumptions of the unimodal adaptation and stasis view of populations.

Sexual Displays: Two Kinds of Divergence under Sexual Selection

Perhaps the best way to begin to shake off the inhibitions that come from an obsession with stability, equilibrium, stasis, and constraints is to think about sex—especially sexual behavior.

Sexual selection is well known to produce some of the most specialized behaviors and morphologies known, like the display and plumage of the male sage grouse, and the antlers of deer. Sexually selected characters show extreme geographic variability, indicated by their divergence in closely related species and subspecies (Darwin 1871; West-Eberhard 1983, 1984). This kind of variation is perfectly compatible with the branching or speciation hypothesis for the evolution of diversity, and with the idea that divergence is associated with reproductive isolation (Mayr 1963).

But there are other kinds of variation that do not fit the branching hypothesis, because they are completely sympatric, intrapopulational, and even "intragenomic." These are the variants represented by "alternative tactics" or polyphenisms. When Darwin (1871) wrote of the "eminent variability" of sexually selected traits, he was referring both to geographic variation and to polymorphisms in the sexually selected sex (usually the males). As an example, consider the dimorphic male beetles of the genus *Podischnus* (Eberhard 1979, 1982). They are strikingly different forms, differing not only in morphology but even more dramatically in behavior. The large, horned males are residential and territorial fighters with specialized dueling, lifting, and dislodging behavior toward rivals. The small males of the same species are dispersers that obtain mates by searching rather than by sedentary territorial defense. Individual form is size-related, apparently influenced by differences in larval feeding.

Here we have a departure from the idea of canalization in a single direction, and stabilizing selection for a single modal form, for this means that divergent phenotypes have evolved in a single life stage without breeding isolation.

For decades theoretical biologists wrestled mightily to explain such complex polymorphism as a special problem. Many have assumed that a genetic polymorphism must be involved and have elaborated models of heterosis (or balancing selection), fitness equilibria, and sympatric disruptive selection to explain this sort of variation (e.g., see Ford 1961; Gadgil 1972; Maynard Smith 1976). This work gives the impression that adaptive polymorphisms require special and unusual conditions for their evolution. This, again, has reinforced the notion that extensive character divergence usually requires reproductive isolation or speciation.

I believe that theory (on the evolution of dimorphism) is off the track in an important way if it fails to consider the developmental nature of these phenotypes (e.g., see Levins 1961, 1968, for early models that do so). In fact, intraspecific alternative phenotypes represent a good place for whole-organism biologists to begin thinking about development in relation to evolution.

Anyone who thinks even superficially about development and behavior has to realize that genetic isolation is not required for phenotype divergence. We see evidence of this in the development of every individual: tadpoles and adult frogs, for example, are strikingly divergent phenotypes produced by the same genotype. The wasps I study, to cite another example, have a winged, long-legged chitinized adult female that lives in a complex world of foraging, nest-building, nest defense, and dominance struggles with her nestmates. As a larva, the same wasp with the same genotype was a wingless, legless, grublike creature that lived a sedentary life in a single cell of the nest, where she was specialized to solicit and process food dispersed by adults. The (human) larva that is most familiar to us also lives a dependent life in a world similarly dominated by trophic concerns and striving for adult attention. Then it gradually grows and changes until it reaches that hormonally mediated metamorphosis we call adolescence, when it transforms into the adult morph, in the male a hairy-faced and relatively independent form. Again we see that one genotype produces two very distinct phenotypes, obviously with no reproductive isolation between the forms.

Some organisms, like host-alternating parasites and "hypermetamorphic" beetles (e.g., Meloidae) go through many discrete stages or forms, each stage with a distinctive set of behaviors and a distinctive "niche." If you consider that complex suites of behavioral traits are also evolved aspects of the phenotype, the number of phenotype variants expressed by a single genotype becomes truly enormous.

You may think that this is all very obvious—and it is. The remarkable thing is that evolutionary theory has never assimilated the significance (for natural selection) of this ubiquitous intraspecific and *intragenotypic* diversity in behavior and development. If it had, we could not seriously consider that a shift to a "new adaptive peak" requires a difficult genetic revolution in a cohesive co-adapted gene pool; or that polymodal adaptation should be rare, requiring special explanation. Intraspecific diversity of specialization is a phenomenon that characterizes normal development and behavior in every form of life (for examples from viruses to higher plants and animals, see West-Eberhard 1986, 1989).

The Genetic and Regulatory Architecture of Alternatives

To understand how genetic evolution can proceed in several directions at once to produce developmental and behavioral diversity within the same species, it helps to consider the regulatory-genetic architecture of flexible traits, especially those controlled by a "switch."

Developmental and behavioral alternatives in general seem to have a common basic structure. This has been demonstrated especially in studies of Batesian mimicry in butterflies (reviewed in Turner 1977) and in studies of other insect polymorphisms (see Nijhout and Wheeler 1982), and is confirmed by studies of the mechanisms of regulation in general (e.g., see Raff and Kaufman 1983; John and Miklos 1988). The genetic-regulatory architecture of a switch is represented in the diagram of figure 4.1. Underlying alternative phenotypes A and B are two sets of genes whose expression (or gene-product use) is controlled by a switch or neurohormonal regulatory mechanism.

The two suites of phenotypic traits are "alternatives" in that they are not simultaneously deployed: either one or the other is expressed, or brought into play, depending on the state of the switch. There is another set of "background" genes or traits not affected by the switch, and expressed in both alternatives. These are sometimes called "nonspecific modifiers" of the alternatives (Turner 1977), in contrast to the "specific modifiers" controlled by the switch. For example, the alternatives "running" versus "walking" would both require the use of legs. So modifiers of leg morphology would be "background" or "nonspecific modifiers" of running and walking behavior. The regulatory mecha-

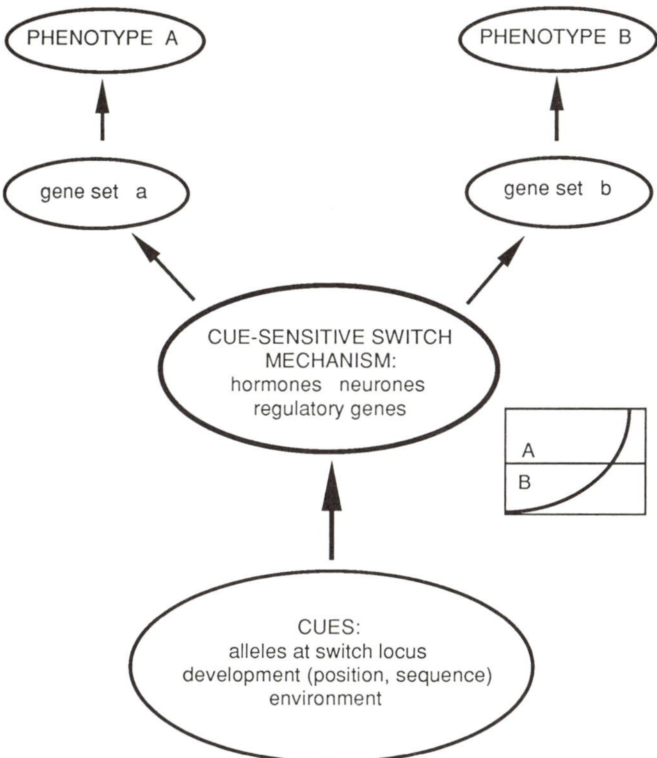

Figure 4.1. The genetic-regulatory architecture of alternative phenotypes. Alternative phenotypes A, B are genetically distinctive in being underlain by different sets of "specific modifiers" whose expression is coordinated by a switch mechanism. Their divergence under selection is to some degree limited ("semi-independent" rather than independent) by shared background alleles, or "nonspecific modifiers" expressed in both alternatives (not controlled by the switch) and essential to their performance. The coordinating or "switch" mechanism (neuronal, hormonal, or regulatory gene product) responds to environmental, developmental, or genetic cues (or some combination of these) in such a way that alternative sets (A, B) of genes or gene products are activated (expressed, or used). Continuous variation in the cues and/or in the response of the coordinating system may be translated into an either-or (switched) phenotypic response by threshold effects (in the coordinating mechanism, or in the response of the regulated modifiers).

nism consists of (1) a sensory apparatus sensitive to environmental cues, and (2) a neuroendocrine coordinating portion, which activates (3) responsive elements of the organism, stimulating gene expression, muscle action, and so forth. The cue can be an environmental factor (like photoperiod, temperature, humidity, diet, or presence of a host, predator, or competitor). Or it can be an internal developmental circumstance, like size or diet at a certain stage; or a genetic factor, like a particular alternative allele or set of alleles at a locus that might then be called "regulatory" locus with respect to a particular set of alternatives. Most commonly a *combination* of genotypic and environmental factors influences phenotype determination, with genetic variance in regulatory elements contributing to the heritability of the response (West-Eberhard 1989).

Note that the alternatives are not depicted as controlled by a linked set of alleles, or a "supergene." It was formerly believed that polymorphisms like the Batesian mimicry morphs of butterflies were both modified and regulated or switched by linked sets of genes on the same chromosome, and that evolution would always tend to produce such linkage (an idea put forth by Fisher 1930). Linked sets of functionally related loci do, of course, occur. Coordinated expression, however, including that of the Batesian mimic morphs of butterflies, now appears commonly achieved by epigenetic or developmental means, not by genetic linkage (see Turner 1977; Raff and Kaufman 1983; John and Miklos 1988). Even if a switch is allelically cued, as in the butterflies, the "switch" (or cue) alleles are now known to constitute a single locus or small set of genes that influence others not necessarily on the same chromosome. It is important to realize, then, that disruptive and frequency-dependent selection, rather than producing two genetically distinct subpopulations or morphs, can lead instead to the evolution of developmentally divergent forms controlled by a regulatory mechanism like that depicted in figure 4.1. This structure permits short-term (phenotypic) plasticity and promotes long-term (genetic, evolutionary) flexibility by not restricting recombination as linkage would (see Thoday 1955).

The regulation of alternatives can be exquisitely condition-sensitive and reversible. But it would be a mistake to consider it "nongenetic" (Hazel et al. 1990). There are three genetically variable points where selection can mold the evolution of facultative alternatives: at loci affecting sensory and response thresholds; at loci affecting the probability of passing a threshold (for example, affecting size in a population with a size-dependent switch); and at loci modifying the alternative phenotypes themselves (the "specific" and "nonspecific modifiers" of the traits.

Switches of this type are extremely common and important in the organization of behavior and related morphology. There is a burgeoning literature to describe "alternative tactics" or "alternative strategies" of resource acquisition, involving mating behavior, foraging behavior, and predation (e.g., see Krebs and Davies 1984). Helping versus reproduction in insects, mammals, and birds, for example, involves switches or "decisions" of this kind. What may be a simple chemical or positional cue in a developmental decision may, in these behavioral alternatives, involve highly refined systems of assessment and learning, and what I think can fairly be called "judgment" regarding decisions between alternatives. In some cases the decisions are fine-tuned by selection to an amazing degree, with the result that alternatives are "adaptive" in the sense of being performed precisely when likely to be advantageous (see Noonan 1981 on kin-directed altruism in wasps).

How can a setup like this possibly evolve? The architecture of a switch and a set of adaptive alternatives would seem to require the simultaneous evolution of adaptive novelty and all this apparatus for condition-appropriate expression. Isn't this an improbable event?

It may seem improbable if you think of evolution as occurring only by small phenotypic steps (strict phenotypic gradualism), with many small genetically specified changes required to produce a large or complex phenotypic change. I believe that we need to think of evolution in a different way: first we should ask how one complex phenotype gets transformed into another; then we can ask what kinds of allelic change might bring this about. If we want to understand how a phenotype evolved, we have to understand how its regulation is organized, and how it might have been pieced together during evolution from preexisting traits. Otherwise we risk imagining kinds of allelic change and even kinds of difficulties that have not occurred.

The Origin of a Behavioral Option: The Worker Phenotype in the Social Hymenoptera

It appears that many condition-sensitive, novel behavioral alternatives are derived from preexisting flexible phenotypes, developmentally rearranged. I can provide an example by going back to the safe ground of my favorite group, the wasps. Most aculeate wasps are "solitary" in that females nest alone, building and provisioning their own brood cells. Those that progressively provision one cell at a time have relatively small ovaries that produce one large egg at a time (e.g., see Evans 1966), alternating with a relatively "ovary empty" phase when that egg is laid. Parallel to the ovarian cycle is a behavioral cycle of switches between cell building, oviposition, provisioning or brood care, and cell closure. Then the female builds a new cell and the two parallel cycles are repeated.

In the taxonomic families of mainly "solitary" wasps, there are a few species that live in groups and share nests. Once I had the good luck of discovering a nest of a rare but famous group-living species called *Zethus miniatus*. This wasp has long been thought to represent a kind of "missing link" between solitary life and the highly social organization of the species that have queens and sterile workers. Its natural history was described in a classic paper published by Adolfo Ducke (1914), who studied it near the Museu Goeldi in Santarem, Brazil. I studied a colony in Cali, Colombia (see West-Eberhard 1987).

Usually, wasps of this species act like solitary wasps: each female builds and provisions her own cells one at a time and defends her own eggs and growing larvae. But they have some striking behaviors associated with the presence of others at the nest: when a female leaves to forage, a neighbor may sneak over and steal food from her cell. But more interesting for the evolution of worker behavior is the fact that females not attending cells sometimes "adopt" and feed orphaned larvae encountered in the nest. Furthermore, females reuse old cells after the brood emerges, and they sometimes fight over empty cells or usurp recently occupied cells from other females and eat their eggs. When this happens, the defeated female leaves the group. She reverts to solitary nesting, the cost of social life with such nasty competitors having evidently proven so high that it is more profitable to endure the tribulations of solitary life (West-Eberhard 1987).

What would it take to make a wasp like *Zethus miniatus* into a "highly social" species with a sterile worker caste? Some theoretical discussions, for example those reviewed by Michod (1982) (see also Parker 1989) depict the spread of a gene for altruism, that is, an allele having a net negative effect on the fitness of individuals in which it is expressed. But it seems to me that selection for selfish obligatory group life could do the same thing: if the solitary-nesting option were closed, dominated females might stay in the group; and, if unable to lay eggs due to social subordinance, their ovaries would regress and they would be held in the "ovary-undeveloped" phase of their cycle. Ovary regression characterizes frustrated or delayed oviposition in many parasitic and aculeate Hymenoptera (Flanders 1969; Pardi 1948). This would produce an altered cycle, with the ovary-developed phase behavior expressed in the "queen," and the ovary-undeveloped phase behaviors expressed in the "worker." The solitary cycle, then, becomes decoupled at a switch between phases, to form complementary "castes" (worker and queen) (West-Eberhard 1987). Note that this would require no gene for altruism—only selection for selfish aggressiveness, brood care (which can be "misdirected"), and group life. Kin selection could play a role in maintaining misdirected brood care among relatives—in making this a positive, rather than a negative, side effect or life in groups. Nor would it require *de novo* evolution of conditional expression of worker behavior: the same cues (e.g., the presence of a larva) that stimulated cyclic brood care in ancestral females could function in altered circumstances (in a group) to stimulate brood care in a non-egg-layer. The cycle has been dissociated into two parts. And this has created two new divergent phenotypes, worker and queen, whose condition-sensitivity was derived from that of the old, ancestral condition-sensitive cycle of reproductive behavior and physiology.

If this hypothesis for the origin of the worker phenotype is correct, it might be possible to create workers artificially simply by forcing solitary females to live in groups. Exactly this experiment has been done in Japan by Sakagami and Maeta (1987a,b) on solitary bees of the genus *Ceratina*. They found that subordinate females in artificially constituted groups did turn into nonovipositioning helpers in the nest.

The Evolutionary Significance of a Switch

All of this is working up to an important point, which concerns the twofold evolutionary significance of a switch. A switch promotes intragenotypic divergence; and it permits dissociation of complex phenotype subunits, or suites of characters, during the course of evolution.

I have just given an example showing dissociation of cyclic alternatives. Consider now how a switch promotes divergence: by coordinating phenotype expression, a switch can link several traits into a developmental and functional subunit, so that selection operates on them as a set. Alternative sets expressed or used in different circumstances can then evolve in different directions. If the regulated forms are "reversible," for example, if alternative behaviors are expressed in a single individual, the divergence is primarily behavioral and physiological, although a surprising amount of morphological divergence may occur, as in the seasonal change in gut length with diet in rock ptarmigan (Gasaway 1976). If the forms are not reversible, but instead irreversibly characterize different individuals, then morphology, or "developmental conversion" (Smith-Gill 1983), may be involved. This has occurred in the three separate lineages of Hymenoptera—the wasps, ants, and bees—that have evolved sterile workers, and in queens. Queens become behaviorally and morphologically specialized to egg-laying; and functional workers become irreversibly sterile females with specialized behaviors of their own, sometimes (e.g., in some ants) lacking morphological equipment, such as developed ovaries and a spermatheca, for independent reproduction.

The same principle applies to the morphs of male beetles. The switch divides the beetle population into two major groups, the big-horned ones and the small-horned ones. There are some intermediates, and of course selection acts on all of them. But selection is most effective at the modes, where it has a larger population of variants on which to act as well as more contrasting initial phenotypes affecting the outcome.

This means that when there are two modes, selection is proceeding simultaneously in what amounts to two different worlds: in this case, the world of the small, hornless beetles and the world of the large, horned ones. The divergence of the two is all the more accentuated if (as in this example) the switch is condition-sensitive. Then not only are there two directions, but they are matched consistently to two different contrasting circumstances. This would not be true of a genotypically controlled switch, which would be condition-blind. Environmental consistency, a consequence of a reliable cue matching phenotype and environment, would speed divergent specialization in tasks performed as alternatives. In effect, selection is "focused" in two different directions.

Recalling the architecture of alternatives (figure 4.1), this means that evolution can proceed in two directions, represented by the two regulated sets of modifier genes (A and B). Clearly this is genetic evolution, as there are genetic differences underlying the two different specialized phenotypes: selection may act differently on variation among large males and on variation among small males—for example, favoring improved horns and fighting ability above a certain size threshold and dispersal behavior below it. In this way more than one "adaptive peak" can be invaded simultaneously by individuals of the same population and life stage. Similarly, genetic divergence can occur between the expressed phenotypes of queen and worker, male and female, or central and satellite males at a lek, even though both divergent sets of genes are carried by every individual in the population.

It is important to realize that the same principle of semi-independent evolution operates in the case of reversible or cyclic behaviors expressed in a single individual. For example, when a wasp is hunting for caterpillars and processing meat, one set of gene products is brought into play; and when she is hunting for wood pulp and applying it as building material to the nest, another only partially overlapping set of traits genetically specified at some level is brought into play. I would regard these two suites of behaviors as evolved—semi-independently evolved, and independently expressed.

I will now dare to propose this as a "law of nature" called the "Rule of Independent Selection." It states that *to the degree that traits are independently expressed they are independently subject to selection.* That is, providing there is genetic variation in each of the traits, they can evolve in different directions.

Thus, once a switch evolves, divergent specialization of the forms separated, or controlled, by the switch can occur. It is important to note here that the switch need not be "genetic," or allelically cued for this to be true. As long as a bimodal (or multimodal) distribution of forms is produced, for example by some recurrent environmental factor, selection operates differently on the different modes and can cause a different set of modifiers to become associated with each.

In some cases, different ecological niches are exploited by different alternative phenotypes. For example, some fish show one kind of locomotory behavior and respiratory mechanism in the water, and another on land during droughts. I would argue that contrasting alternatives of this kind are a major source of "macroevolutionary" change: rather than requiring a series of speciations, such change can occur with no speciation at all, due to a behavioral or developmental switch, under the "rule" of independent elaboration of independently expressed forms. The vertebrate invasion of the land almost certainly got started as an alternative specialization (see Graham et al. 1978).

Alternatives and the Origin of Major New Traits

Divergence between intraspecific alternatives may prove to be a more important source of major evolutionary change and entry into "new niches'' than divergence of geographic isolates. This expectation is based on two qualities of polyphenisms. First, there is the buffering effect of alternatives. If the fish population that first invaded land had been monomorphically adapted so that it had to change gradually to track an environment with more and more frequent droughts, it would presumably go through a transitional period of being intermediate and poorly adapted both in water and on land. The developmental trick of a switch between specializations enables a new suite of traits to be elaborated without loss of the potential to express the old. The population never has to pass through the ''deep valleys between adaptive peaks,'' which is seen by conventional theory to accompany major evolutionary change, whether under selection in changing environments (Fisher) or due to drift (Wright) (see Futuyma and Slatkin 1983). There is no ''genetic revolution'' while an ancestral phenotype is reorganized to produce a novelty. The polyphenic fish specializes to land without despecializing to water.

The second reason for the possibly great importance of alternative phenotypes for macroevolution is what I call the phenomenon of ''oppositeness'' (or ''antagonism'') of alternatives (West-Eberhard 1979), which may accentuate divergence. In the evolution of alternatives there can be a premium on divergence per se, to put a new option outside the competitive realm of an established one. Selection for oppositeness can then continue to drive the two options in different directions. Here is an example (from Eberhard 1980): in weevils of the genus *Rhinostomus* (*R. barbirostrus*) large males defend females. They fight by violent flipping at each other with an enormous noselike rostrum; the rostrum is not an olfactory device, but a weapon. Small males have little chance of winning in such a contest. They adopt the alternative behavior of lurking near guarded females and sneaking copulations. Sneaking, which entails behaviors opposite to fighting and threat, in terms of conspicuousness and strength, sometimes works for small individuals; and it is a tactic that cannot be played so effectively by the more cumbersome large phenotype. It seems likely that selection for ''oppositeness'' has made the exceedingly small ''minor'' morph of this beetle genetically small by favoring specific modifiers for exaggeration of small size. Although the genetic basis to this variation remains to be investigated, selection for extreme adaptive divergence in size of morphs is illustrated by the evolution of extreme sexual dimorphism in size in some species (see Ghiselin 1974).

Selection for oppositeness may have played a role in the evolution of anisogamy, or sexually dimorphic gametes—the ultimate basis for all behavioral and structural sexual dimorphism. Disruptive selection apparently led, on the one hand, to large eggs, specializing in a sedentary nutritive tactic; and, on the other hand, to small sperm, specializing in motility, dispersal, and competitive searching (see Parker et al. 1972; West-Eberhard 1979). Out of that basic divergence came the sexual dimorphism of female and male, backing up the dimorphism of their opposite gametes. Sexual dimorphism has the same basic structure just described for alternative phenotypes in general (figure 4.1) and is subject to the same evolutionary rules of semi-independent selection and divergence.

The Modular Organization of Phenotypes

For visualizing the switch-controlled phenotypes that I think are especially important for diversifying evolution, there is a term invented by Dobzhansky that is especially apt. Dobzhansky (1970) referred to the "phenotype repertoire" of a genotype, a term similar in meaning to "reaction norm" but more suggestive of multiple discrete sets of developmentally "individualized" or semi-autonomous traits (see Wagner 1989). Each genotype has the capacity to produce many phenotype subunits, or sets of co-expressed traits (see also Darwin 1868). Each functions as a unit, and is expressed or employed as a unit, under the control of some regulatory mechanism or set of regulators that controls their expression, integrating what Bonner (1988) has called a "gene net." The repertoire of a genotype may include all of the kinds of semi-independently expressed and evolving phenotype subunits I have been talking about: different life-stage phenotypes; sex-limited character sets, which are generally regulated by hormonal coordination and a small number of sex-determining alleles or a dimunitive chromosome, or even an environmental cue; alternative male behavioral tactics; a female reproductive cycle; and complementary alternative behaviors like hunting and foraging. (I have not included the kinds of switches that determine different tissues and organs; I will discuss these "interdependent" and therefore difficult to dissociate subunits later.)

I realize that this modular or subunit conceptualization of the phenotype is oversimplified and naive as a representation of development and behavior. For example, the switch determining the adult queen-versus-worker phenotypes in some ants occurs way back in the egg stage (see Wheeler 1986), and triggers a cascade of effects not adequately summarized by considering its endpoint, a functionally coordinated set of traits observable in the adult. And there may be "compound" switch mechanisms with a common cue, or position effects not adequately represented by thinking in terms of a "switch." But a concept of phenotype subunits or modules, even if crude, is a way of linking conditional specialization in behavior and development, on the one hand, to selection and genetical evolution, on the other. The "subunits" represent units of both development or gene expression and selection, or adaptive function and evolution: not only do the developmental and behavioral subunits evolve somewhat independently so that they may diverge from each other, but they are dissociable—they can be shifted, lost, or regained as units during the course of evolution (see Darwin 1868).

Evolutionary shifts in timing, or stage-of-expression, of sets of traits—heterochronic changes—are well known in both plants and animals (Lord and Hill 1987; Gould 1977). But I would argue that the dissociation of same-stage alternatives of the kind I have been talking about may be more frequent and more important in evolution. The reason is that alternative phenotypes are expressed as *options*. An alternative behavior pattern can evolve, and be elaborated, alongside an established pattern without having an equal or superior fitness payoff: it need only be superior in certain circumstances. Since it is an option, it can be dropped in changed circumstances without functional disruption of the organism, which is left with an alternative. For a heterochronic shift to spread and persist, it has to be superior to the stage-specific phenotype it replaces. And monomorphic stage-specific phenotypes sometimes may be integral building blocks in ontogeny, such that altered timing disrupts development (discussed in Thomson 1988).

To illustrate evidence for loss of a complex behavioral alternative I will turn again to the behavior of wasps. In Australia, Howard Evans and Robert Matthews (1975) studied some predatory wasps (*Bembex*) that usually hunt flies. Some species were behavioral polyspecialists, stocking their nests with both flies and damselflies, or, in other species, flies and Hymenoptera. Although Evans and Matthews did not describe hunting behavior, this probably involved two divergent sets of traits, because these different kinds of prey behave differently, were found in different habitats, and would require different modes of handling and transport. In addition to the polyspecialists, there are some closely related species that hunt only damselflies and some that specialize in Hymenoptera. Evans and Matthews speculate that these new secondary specializations may have been derived via a polyphenic stage, where they hunted two kinds of prey (e.g., flies and damselflies), followed by what I call ''phenotype fixation'' to perform only one behavior pattern (e.g., hunting damselflies). If this interpretation is correct, the two options have been dissociated by the loss of one.

Suites of co-expressed traits differ in their degree of dissociability, and it is of interest to distinguish ''dependent'' and ''independent'' alternatives. The castes of social insects—specialized workers and queens—are divergent but not dissociable: no one form, in a specialized species of ant, can reproduce on its own. Rather, like the tissues and organs of a multicellular organism, they are mutually dependent on each other; like hands and feet, or liver and lungs, they are not readily dissociable. On the other hand, the aerial and aquatic leaves of heteromorphic plants (e.g., *Ranunculus* species), even though they look like organs of an individual, are ''independent'' dissociable aspects of phenotypes. Growth on land produces a full independent terrestrial form lacking the aquatic leaves; and submerged plants lack the aerial leaves (see Cook and Johnson 1968). I use this example partly to point out that the condition-sensitive growth of plants and other modular organisms is very much like the flexible behavior of animals (see Silvertown and Gordon 1989). So the generalization I am trying to make here seems to apply to plants as well as to animals, and to viruses and bacteria, or to any form of life where alternative phenotypes are to be found.

Conclusion

It is quite clear that we are moving into a new era in evolutionary biology—one where a fundamentally genetic theory is expanding to include development, not just as a ''constraint'' but as an agent of change.

This will require full incorporation of the idea that phenotypic variation is the material basis for selection; that the environment plays a guiding role (alongside the genes) in the construction of phenotypes; and that individuals are mosaics of differentially evolving traits; they are neither monolithic individuals under selection, nor are they dependent entirely on the particulate originality of single mutant genes for the origin of adaptive novelty.

Key evidence for this expanded view of evolution will certainly come from endocrinology, neurobiology, and molecular genetics. But it would be a mistake to believe that all that is interesting and important will happen in those fields. There is a whole new frontier for comparative study aimed at discovering how what I would call ''judgment'' or adaptive decision-making ability is pieced together from ancestral phenotypes. I foresee a new mode of experimentation, in which rudimentary flexible or apparently fixed phenotypes are experimentally “pushed” to produce semblances of complex, derived phenotypes in order to test hypotheses for the origins of adaptive flexibility, along the lines of the experiments by Sakagami and Maeta described above. Ketterson et al. (ms.) call this “phenotypic engineering,” an example being their research on the behavioral and fitness effects of experimentally altered testosterone levels in a field population of juncos (*Junco hyemalis*).

An epigenetic approach will mean asking new kinds of questions: for example, when parental care shifts, in an evolutionary lineage, from one sex to another (as it has done repeatedly in fish, birds, mammals, and frogs) does this require the *de novo* evolution of a new set of behaviors in the newly parental sex? Or is all or part of a suite of ready-made adaptive traits simply transferred between the sexes via evolved hormonal change, as in the case of the famous signaling genitalia of female hyenas and marmosets (Eibl-Eibesfeldt 1970; Lindeque and Skinner 1982); or the capacity to sing, which, complete with associated brain morphology, can be transferred between the sexes in hormonally treated canaries (Nottebohm 1980). What maintains genetic variation in the threshold for adopting different alternative phenotypes, as that documented for performance of different tasks by social insect workers (a question discussed by Page et al. 1989)? Does behavior usually take the lead in the elaboration and determination of complex new traits (Wcislo 1989), or is the evolution and determination of behavioral flexibility often influenced by morphological variation, as occurs in the seed-size specialization of individual Darwin's finches (*Geospiza fortis*) (Grant 1986)? These are the sorts of questions that will become more interesting as “epigenetic” phenomena increasingly attract the attention of evolutionists.

Mainly, I predict a profound change in the ways of thinking about the structure and transformation of phenotypes. Waddington's (1975) classic diagram of the epigenetic landscape, so often reprinted as a portrait of canalization and unimodally directed development, will be seen instead as he intended it—as a diagram of plasticity and latent potential for diversifying evolutionary change.

Summary

Behavior and development are condition-sensitive processes whereby different complex phenotypes are produced by a single genotype. Although it is common to think of populations as unimodally adapted, with speciation required for the origin of divergent novelties, in fact major divergence may often originate as a bifurcation in the pattern of development or behavior, producing intraspecific alternative adaptations. Like life-stage and sexual differences, alternative phenotypes are controlled by a neural or hormonal regulatory mechanism whose condition sensitivity may be genetically variable and subject to evolution. A hypothesis to explain the simultaneous origin of a new adaptive alternative and its condition-sensitive expression is suggested by comparative studies of wasps and bees showing how facultative worker behavior may have evolved. Developmental and behavioral switches divide the phenotype into subunits that evolve semi-independently, according to the "Rule of Independent Selection" of independently expressed traits. Alternatives promote major change due to selection for oppositeness and the buffering effect of multiple options. Switches also render phenotype subunits dissociable, giving rise to heterochronic and transsexual changes, as well as losses and recurrences of complex traits as units. An epigenetic view of the phenotype suggests new emphases for research in evolutionary biology.

Acknowledgments

For useful suggestions I thank Ross Crozier, William Eberhard, Frank Joyce, Ellen Ketterson, Ernst Mayr, George Williams, and the editors. Gillian Holder and Dolores Mills helped with preparation of the manuscript.

5

The Middle Ground of Biology: Themes in the Evolution of Development

LEO W. BUSS AND
MATTHEW DICK

During John Bonner's professional lifetime the biological sciences became increasingly specialized and fractionated. Biology became "big-time"; the cult of the system, the technique, and the laboratory became increasingly paramount. We have become accustomed to an era when it is not simply possible, but likely, to find that an individual with a doctorate in the biological sciences is ignorant of the existence of entire phyla. Throughout this period, John Bonner has been one of the few articulate voices for synthetic thinking in biology. He has repeatedly asked us "why one cannot be a reductionist and a holist at the same time" (Bonner 1988, ix). He has "a hankering for middle-sized questions" (Bonner 1965, p. v), the middle being somewhere between the physical sciences and philosophy, and a generation of students raised on his books have had the "middle-sized questions" posed for them.

This trend is beginning to reverse, in some measure as a consequence of progress in his specialty, developmental biology. Spectacular progress in developmental genetics now invites phylogenetic interpretations and rekindles old debates on the role of development in major evolutionary events. On other fronts, the roles of development in life cycles and life cycles in evolution have touched evolutionary theory as its core assumptions and provide guidance for the next stage in maturation of that ever-robust theory.

We provide vignettes of three such "middle-sized questions" emerging from this new spirit of synthesis, chosen to celebrate the areas of great promise and to identify problems that remain particularly recalcitrant. To the former we add speculations to fuel an already white-hot fire. The latter, those less worthy of celebration, are no less worthy of prominence. On these matters we will attempt to say "familiar, well-known things backwards and inside out, hoping that from some new vantage point the old facts will take on a deeper significance'' (Bonner 1958, p. 1).

The Rebirth of Speculative Zoology

> *Therefore, the idea that evolution and development should be brought together is not new, nor has it been totally neglected in recent years. What is new is that there is suddenly a general consensus that this is precisely what is needed at this time. There is a sentiment that a knowledge of development will give us greater insight into mechanisms of evolution and that a knowledge of evolution will give us corresponding insight into mechanisms of development.*
>
> (Bonner 1982, p. 4)

The great era of speculative zoology, realizing both its zenith and greatest excesses in the work of Haeckel, was all but spent by the turn of the century. Beginning in 1887, with Chabry's experiment on the embryos of an ascidian and followed closely by more widely appreciated works of Roux and Driesch on frogs and echinoderms, respectively, the new field of developmental mechanics was launched. It appeared that differences between embryos, long the subject of phylogenetic speculation, could be understood experimentally. As experiments were repeated with different groups and as patterns began to emerge independent of phylogenetic interpretation, adherents were led to the opinion, voiced so clearly by C. O. Whitman, that ''we have no longer any use for the 'Ahnengalleries' [ancestor portrait gallery] of phylogeny. . . . We are no better off knowing that we have eyes because our ancestors had eyes. If our eyes resemble theirs it is not on account of genealogical connection, but because the molecular germinal basis is developed under similar conditions'' (1895, pp. iii–iv). Developmental biology, the successor to developmental mechanics, grew isolated from evolutionary biology. Indeed, it is well known that developmental biology was all but omitted from the formative era of the development of the Modern Synthesis (Mayr and Provine 1980), an omission that one of us has argued can only be justified by tacit acceptance of the Weismannian dogma (Buss 1987).

Development was no less isolated from genetics during this same era. As Mendelian genetics became established, a fundamental and curious contradiction arose. The chromosome theory held as central that genes were responsible for ontogeny: their faithful replication and equal partitioning were observable in any cell independent of its state of differentiation, and mutations occurring in them were expressed in altered developmental patterns. Yet, developmental biologists had, by this time, over twenty years of experimental evidence that the nucleus was passive in earliest ontogeny; it was the cytoplasm that directed development. Experimental manipulation of cytoplasmic determinants altered development; the nucleus could often be inactivated with no effect. Both sides had seemingly unequivocal experimental evidence in support of their positions. As F. R. Lillie put it, "The progress of genetics and physiology of development can only result in sharper definition of the two fields, and any expectation of their reunion is in my opinion doomed to disappointment" (1927, p. 367). At minimum, Lillie continued, "Those who desire to make genetics the basis of physiology of development will have to explain how an unchanging complex can direct the course of an ordered developmental system" (1927, p. 367). With the advent of molecular biology, these and other problems were rapidly resolved, and a hybrid field, developmental genetics, has come into being with the resolution of Lillie's challenge as its central agenda.

We celebrate not this reunion, but rather the wider and older gulf that has separated the intellectual descendants of Roux and those of Haeckel. The spectacular successes of developmental genetics, particularly of *Drosophila*, invite another visit to the "Ahnengalleries." As the genetic circuitry underlying the establishment of segmental organization in the fly embryo becomes established, new perspectives on old problems are emerging (figure 5.1). Plausible examples abound. For example, deletion of *Antp* reveals segmental designs reminiscent of that of apterygotes, deletion of the entire BX-C reproduces aspects of the segmental design of myriapods, and deletion of both the BX-C and ANT-C generates the segmental design of onychophorans. Such patterns have been interpreted as atavisms (Raff and Kaufman 1983). Similarly, the paired segments of millipedes and centipedes have been hypothesized to result from different patterns of regulation of pair-rule genes (Sander 1988; Minelli and Bortoletto 1988). To this list, we add an additional example.

A wide variety of polychaetous annelids are capable of asexual reproduction (figure 5.2), and the capacity to reproduce in this fashion has probably evolved repeatedly. While virtually any of these forms could bear reexamination with the recent knowledge of *Drosophila* embryo-

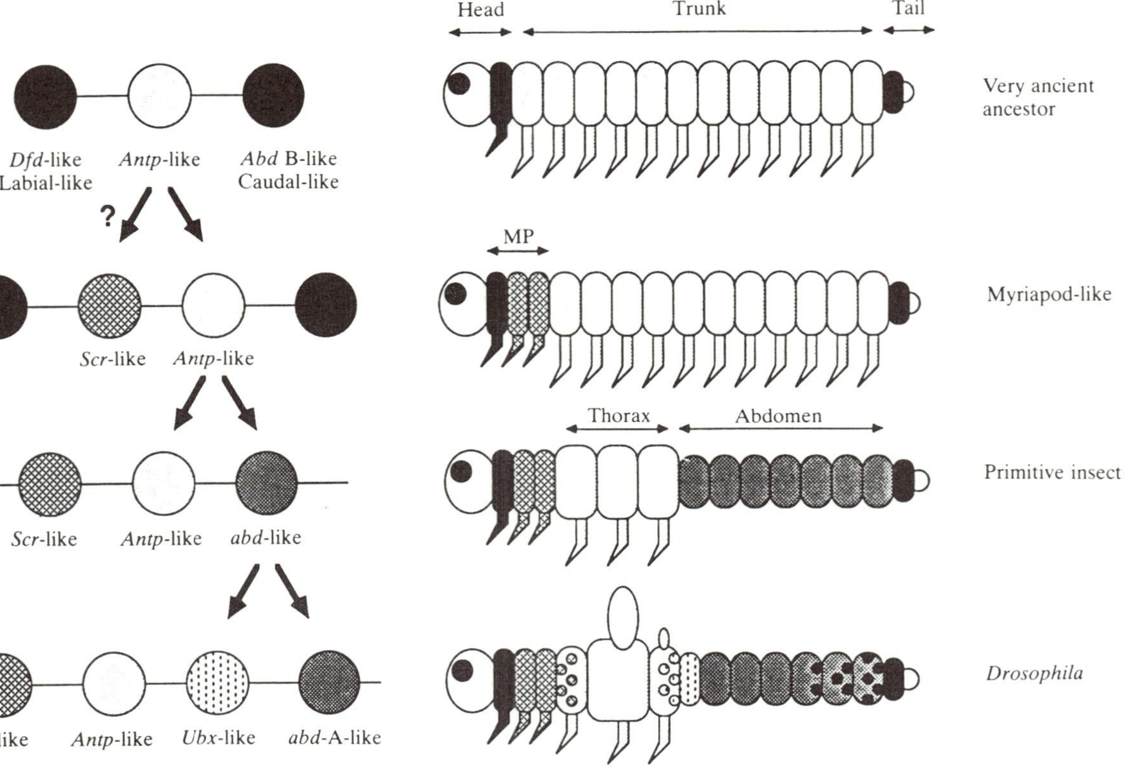

Figure 5.1. Scheme for the origin of homeotic genes during evolution of the myriapod-insect lineage. Panel on left shows the proposed set of *Antp*-like genes existing at each stage in the evolution of the lineage. Black circles (shown only for stages 1 and 2) indicate the existence of other homeobox genes that probably arose prior to the isolation of this lineage. Arrows indicate potential gene duplication events. Corresponding shading on the diagrams at right shows the proposed expression domain for each gene, signifying progressive specialization of trunk segments. Complex patterns of gene expression in the head and tail are depicted only by black boxes. (From Akam et al. 1988.)

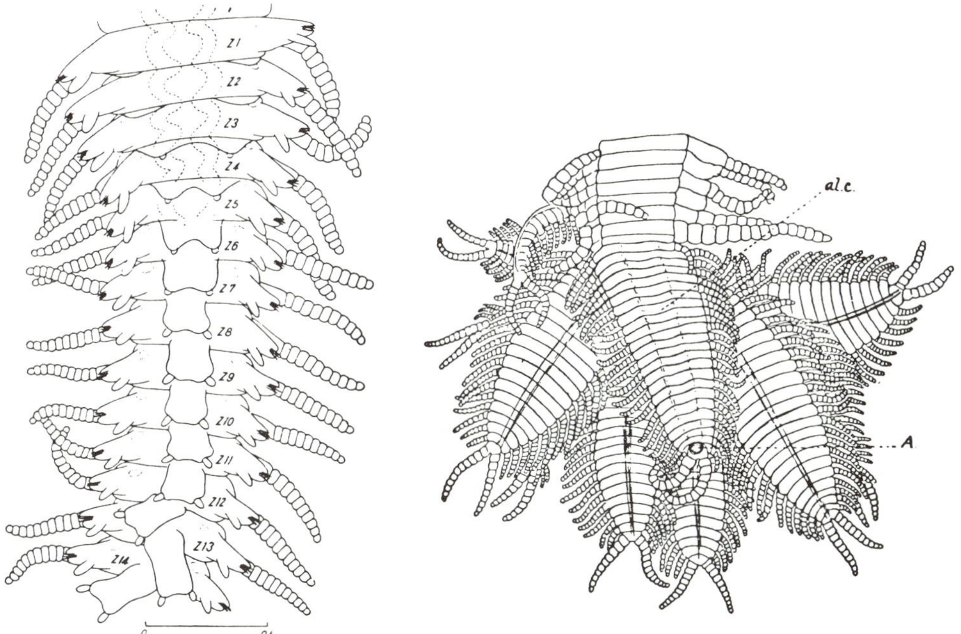

Figure 5.2. Modes of asexual reproduction in syllid polychaetes. *Left*: Mode of stolonization in *Trypanosyllis asterobia*, showing a line of stolons, or differentiating segments of the parent worm, with different degrees of development of pygidia in a series from anterior (Z1) to posterior (Z14). *Right*: Caudal buds of *Trypanosyllis gemmipara*, with parent worm at top; *al.c.*, posterior ciliated portion of intestine; *A.*, anus. (Left from Okada 1937; right from Johnson, 1902.)

genesis in mind, the annelid *Ctenodrilus serratus* is exemplary. *Ctenodrilus* is a small organism, roughly 2 mm in length, composed of fewer than twenty segments. It lives in mud, feeds on detritus, and is an undistinguished and little-studied component of the North Sea meiofauna. Its mode of asexual reproduction is nothing short of fantastic.

The worm undergoes what is known, in the complex terminology governing modes of annelid reproduction, as "paratomic fission" (figure 5.3). Fragments are serially produced along the body of the "parent" worm and differentiate into more or less complete organisms prior to separation (Peters 1923). What is remarkable is that aspects of the subdivision and manner of differentiation of the worm body during paratomic fission are reminiscent of the sequence of expression patterns of developmental regulatory genes observed in *Drosophila* during embryonic pattern formation. The worm body consists of anterior and posterior regions of four to five segments that do not fragment, and a central

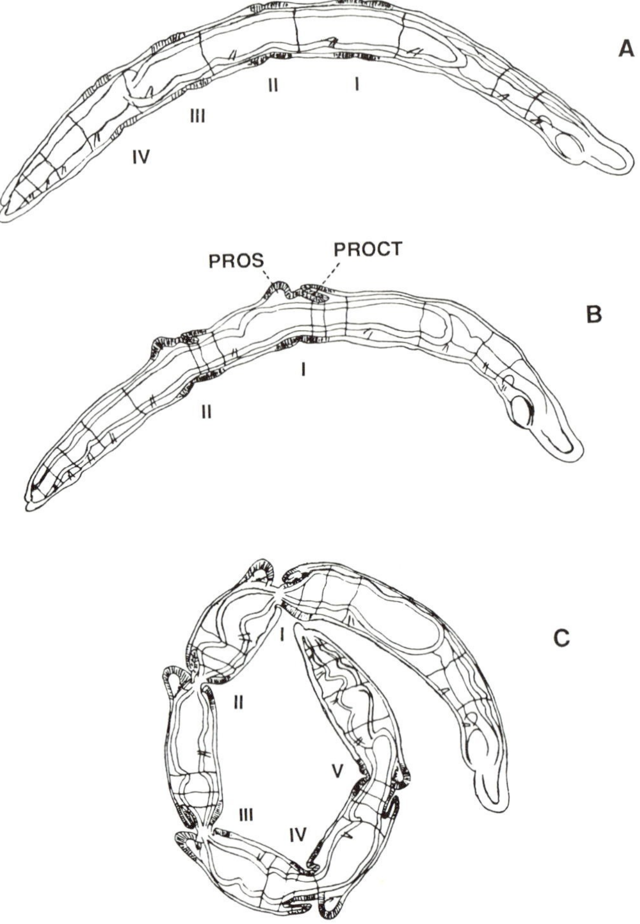

Figure 5.3. Paratomic fission in *Ctenodrilus serratus*. (A) Worm with four separation zones (I–IV), in early stage of fission process. (B) Worm with two separation zones showing proctodeal (PROCT) invaginations and prostomial (PROS) rudiments. (C) Worm with five separation zones nearing fragmentation. (From Peters 1923.)

region of one to nine segments that do (figure 5.3a). These regions are reminiscent of broad domains established by gap genes in *Drosophila* development. Separation zones become apparent as thickenings of the body wall around slight annular constrictions, and occur between septal boundaries. A fragment thus includes a septum and part of the segment on each side of it. Separation zones established between septal boundaries are similar to parasegmental zones established in *Drosophila* by complementary domains of some pair-rule genes (see Akam 1987 and Ingham 1988 for pertinent reviews), and indeed the portion of a parent worm that contributes to a new individual corresponds directly to a parasegment. Each fragment develops a prostomium and a pygidium from the dorsal surface of the thickened anterior and posterior boundaries, respectively (figure 5.3b). The worm parasegments are polarized; terminal structures develop facing one another on either side of each separation zone, suggestive of the expression domains of segment polarity genes. Two new partial transverse septa arise in the proctodeal area but are not completed, nor do setae develop, until after fragmentation (figure 5.3c). Homeotic genes are implicit in changes in cell fate from the relatively uniform structure of the parent parasegment to the head, posterior segments, and tail of the new individual.

The phenomenological match between patterns of *Ctenodrilus* paratomic fission and *Drosophila* early development is not an isolated example. The process of delineating specific body regions, reminiscent of the sequelae of gap gene expression, occurs not only in annelid asexual reproduction, but is also seen in regeneration (Okada 1929; Berrill 1931, 1952). Indeed, regenerating annelids display several features reminiscent of patterns of segmentation in *Drosophila* mutants, including segment skipping, mirror-image duplication (e.g., figure 5.4, pygidium formation at anterior ends in *Autolytus edwardsi*), and ectopic duplications of structures (e.g., figure 5.5, sequential duplication of heads in *Autolytus pictus*).

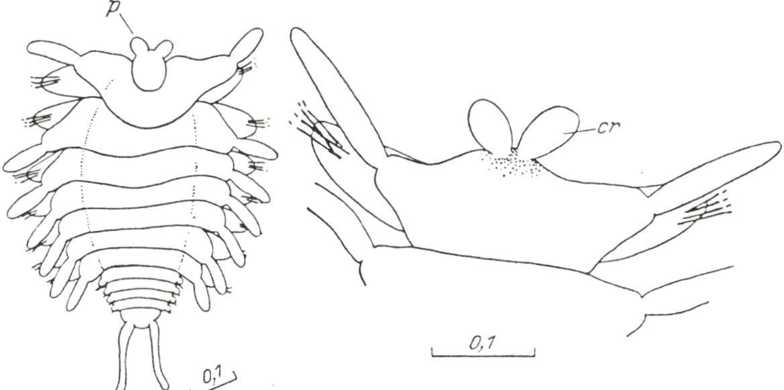

Figure 5.4. Tail regeneration at anterior end of *Autolytus edwarsi*. *Left*: Regeneration of whole pygidium. *Right*: Regeneration of caudal cirri alone. (From Okada 1929.)

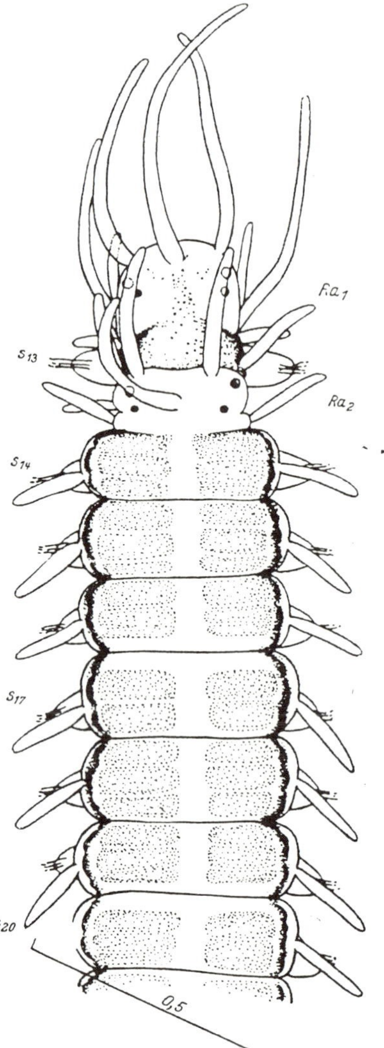

Figure 5.5. Double head regeneration in *Autolytus pictus*. (From Okada 1929.)

Whether similar phenomenology implies similar mechanisms of genetic control in these organisms is clearly an unresolved empirical question. To the extent that some mapping does indeed exist, the fact that the temporal progression of gene expression in *Drosophila* embryogenesis proceeds from gap to pair rule to segment polarity to homeotic selector genes, and that an analogous spatial and temporal progression appears in fission by *Ctenodrilus*, is of particular interest. These similarities suggest the possibility that not only are homologs of genes from several levels of the hierarchical network regulating *Drosophila* development present in *Ctenodrilus*, but that their original role in embryonic pattern formation has been exapted to perform an additional role in asexual propagation, which occurs late in ontogeny. There is no conceptual reason why developmental regulatory cascades cannot be turned on at times in ontogeny other than embryonic development by different internal signals (e.g., hormonal influences on annelid reproduction and regeneration are well known; see Clark 1965, 1966, 1969; Clark and Olive 1973; Franke and Pfannenstiel 1984). The descendants of Roux and Haeckel have something to talk about after all.

Seeing the Possible

> *Let us consider the problem of how complexity of an organism affects its ability to change. On the surface it would seem obvious that if an organism is small and has relatively few parts and few genes, it would be easy to affect by mutation and recombination, the kind of changes needed for evolution.*
>
> (Bonner 1988, p. 166)

Throughout the history of post-Darwinian biology, there has been a tension that recurs at regular intervals, a problem that manages to insert itself into whatever the debate of the day might be, a problem that remains unresolved. Simply stated, the problem is this: while most biologists are content that our understanding of evolutionary processes is inadequate to accommodate patterns of change displayed within species and within lineages of lower taxonomic rank, there has consistently been doubt in some quarters that these same processes are adequate to understand the origin of higher groups.

Francis Galton, Darwin's cousin, was among the first to raise doubts, advocating that "the theory of Natural Selection might dispense with a restriction, for which it is difficult to see either the need or the justification, namely, that the course of evolution always proceeds by steps that are severally minute, and that become effective only through accumulation" (1889, p. 32). T. H. Huxley, "Darwin's bulldog," differed on this point with Darwin no less, preferring to view gaps in the fossil record as evidence for "saltation" rather than an artifact of the vicissitudes of deposition. The debate returned with particular force at the turn of the century in the context of the battle between Mendelians and biometricians (Provine 1971). William Bateson, whose stridency on these matters has not often been equaled, voiced the Mendelian and mutationist position that to the degree to which "the process of Evolution is found to be discontinuous the necessity for supposing each structure to have been gradually modelled under the influence of Natural Science is lessened, and a way is suggested by which it may be found possible to escape from one cardinal difficulty in the comprehension of Evolution by Natural Selection" (1891, in Punnett 1928, p. 128). As Fisher and others came to accommodate an attenuated form of mutationalism in their formulation of the synthetic theory of evolution, the passion of earlier days subsided. The "one cardinal difficulty" refused to remain long submerged. Goldschmidt (1940), in particular, claimed that "if one tries to work out this idea in detail one soon comes to a point where it is evident that something besides Neo-Darwinian tenets is needed to explain such processes." Goldschmidt's "hopeful monsters" recall de Vries's mutation theory. The notion that something special accompanies the origin of major groups has reappeared recently. Gould (1977), for example, was led to "predict that during the next decade, Goldschmidt (1940) will be largely vindicated in the new world of evolutionary biology." Even more recently, the discussion has reemerged in the context of the units-of-selection debate (Buss 1987). Doubters aside, the orthodoxy has reigned throughout the history of this debate and remains vigorously defended today (Bock 1979; Lande 1980; Charlesworth et al. 1982; Wallace 1985; Levinton 1988; Hoffman 1989). Evolution proceeds via small steps; large steps are but the accumulation of small steps, and small steps are directly observable.

We offer no resolution to this problem but wish to note that a class of data that bears directly on the issue has yet to be explicitly considered in this context. The problem, stripped to its essentials, is the question of what is possible. We need to see the entire range of developmentally possible variation exhibited by some complex eukaryote that is attainable in some small number of generations. Techniques are available to do so. Developmental geneticists routinely apply saturation mutagenesis to retrieve mutants of a desired phenotype. Saturation mutagenesis should, in principle, reveal the possible.

The products of mutant screens of *Drosophila* are, in fact, dramatic. Among the most striking mutants are the homeotic variants. While the segments of homeotic mutants are undeniably drosophilid in character, their arrangement clearly is not. Several have read into these variants the history of higher taxa (Raff and Kaufman 1983). However, so far saturation mutagenesis has failed to produce anything vaguely prospective.

Would we see something different if developmental geneticists had chosen to saturation-mutagenize some form that was basal to subsequent radiations? Sewall Wright (1982) believed this to be the case (figure 5.6), stating:

> Such changes would probably have been impossible except in an organism of very small size and simple anatomy. I have recorded more than 100,000 newborn guinea pigs and have seen hundreds of monsters of diverse sorts (Wright 1960) but none were remotely "hopeful," all having died shortly after birth if not earlier. Yet among nine specimens of a small trematode (*Microphallus opacus*) about 1.5 mm long, of which I made serial sections in my first research project (Wright 1912), one was highly abnormal in form and had two large ovaries instead of only one. It would probably have been considered a monster if it had been a large complicated organism, but it was apparently flourishing as well as the others before it was fixed. (p. 440)

On this basis, Wright (1982) comes to the conclusion that

> It may seem that mutations with impossibly drastic effects would be required for the origins of the higher of the taxa. Such origins, how-

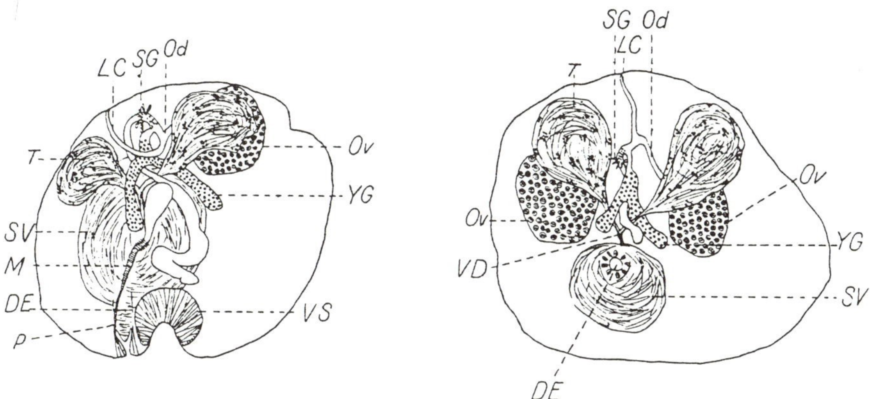

Figure 5.6. Reconstruction from transverse sections showing arrangement of reproductive organs (except middle of uterus) in the trematode *Microphallus opacus*. *Left*: Typical individual with a single ovary, Ov. *Right*: Abnormal individual with two ovaries. DE, ejaculatory duct; LC, Laurer's canal; M, metraterm; Od, oviduct; P, copulatory papilla; SG, so-called shell gland; SV, seminal vesicle; T, testis; SV (= VS), seminal vesicle; YG, so-called yolk gland. (From Wright 1912.)

> ever, probably all occurred from species, the individuals of which were so small and simple in their anatomies that mutational changes, that would be complex in a large form, were not actually very complex. (p. 442)

There are at least two ways to interpret this claim. The strong form has a distinct nonuniformitarianist implication. In the past, organisms were simple, and this simplicity permitted great change with little genetic tinkering. Today, even small and seemingly simple forms are buffered from these changes. Hence, we cannot obtain novelty from existing forms; novelty can be experimentally elicited only if the experiment is done in the Cambrian. The weak form, however, is the one that Wright clearly favors. Complex organisms, and here he was clearly referring to morphological complexity with no implication of genomic complexity (see Wright 1982), cannot produce novelty, but simple forms can. The weak form is testable; one would like to see the results of saturation mutagenesis of a turbellarian flatworm.

Problems that won't go away may be real. This particularly recalcitrant problem will only be successfully addressed by critical exposure to primary data. If, as most of us believe, evolution is reducible to understanding heritable variation and selection, then any device that gives us a glimpse of the range of the possible variation provides a central source of data for evolutionary science. Improbable as it may seem, novelty of the sort recognized taxonomically at the level of the genus or higher might be produced. Sequential rounds of saturation mutagenesis and selection either will or will not lead to novelty. It is a curious symptom of the fractionization of the biological sciences that this technical tool of developmental geneticists has not been seen as providing data central to this problem.

Doliolum and Life Cycle Evolution

> *It is an interesting paradox that at all levels of complexity there seem to be forces which bring components together, as in the integration of cells into individual organisms, or gatherings of organisms into social groups; but at the same time there are other forces which work in the opposite direction and cause those individual organisms or those social groups to be isolated from one another. Curiously, the forces which seem to be operating in opposite directions*

> *producing integration and isolation are the same; they are the forces of natural selection.*
>
> (Bonner 1988, p. 229)

Instructors of invertebrate zoology always have means at their disposal to discriminate between those students who merely know the material and those who know the material cold. The favorites vary from instructor to instructor, but virtually all consider the complex life cycles of some invertebrates as material difficult to master. The habits of chondrophores and siphonophores among the hydrozoans and habits of cestode and trematode parasites among the flatworms make perennial appearances on examinations. The life cycle of the thaliacean *Doliolum*, though, is the ultimate discriminator (figure 5.7).

Its life cycle is extraordinary. A sexually produced larva gives rise to a free-swimming form called the oozooid (figure 5.8a). The oozooid, superficially similar to a salp, serves several functions: it feeds, it respires, and it locomotes. It does not, however, reproduce sexually. Rather, the oozooid gives rise to a series of minute prebuds from tissues of a filamentous stolon. The prebuds are technically free-living; indeed, if it were not for a specialized group of amoeboid cells, the phorocytes, they would depart from the mother oozooid. The phorocytes, however, capture the prebuds before they release and transport them, in twos and threes, to a dorsal outgrowth (the cadophore) of the mother (figure 5.8b). The prebuds divide and become attached in medial and lateral positions. Each lateral bud gives rise to a new form, the gastrozooid (figure 5.8c), which remains permanently attached to the oozooid. When gastrozooids become functional, the oozooid loses its gut and branchia; thereafter, gastrozooids take on the feeding and respiratory roles, leaving the oozooid only its locomotory and asexual reproductive functions. The gastrozooids, though, are technically free-living; they settled on the mother only with the aid of phorocytes. Communication between the gastrozooids and the oozooid requires the development of a placentalike structure, through which gases and nutrients are passed by diffusion.

Yet this is hardly the oddest feature of the life cycle. Those buds, derived from the initial prebuds, that settled in clusters in the medial row on the cadophore, have a different fate. One bud of each medial cluster gives rise to a distinct zooid type, the phorozooid (figure 5.8d). Phorozooids are morphologically similar to oozooids and are initially attached to the oozooid by a narrow stalk. Another bud in the cluster

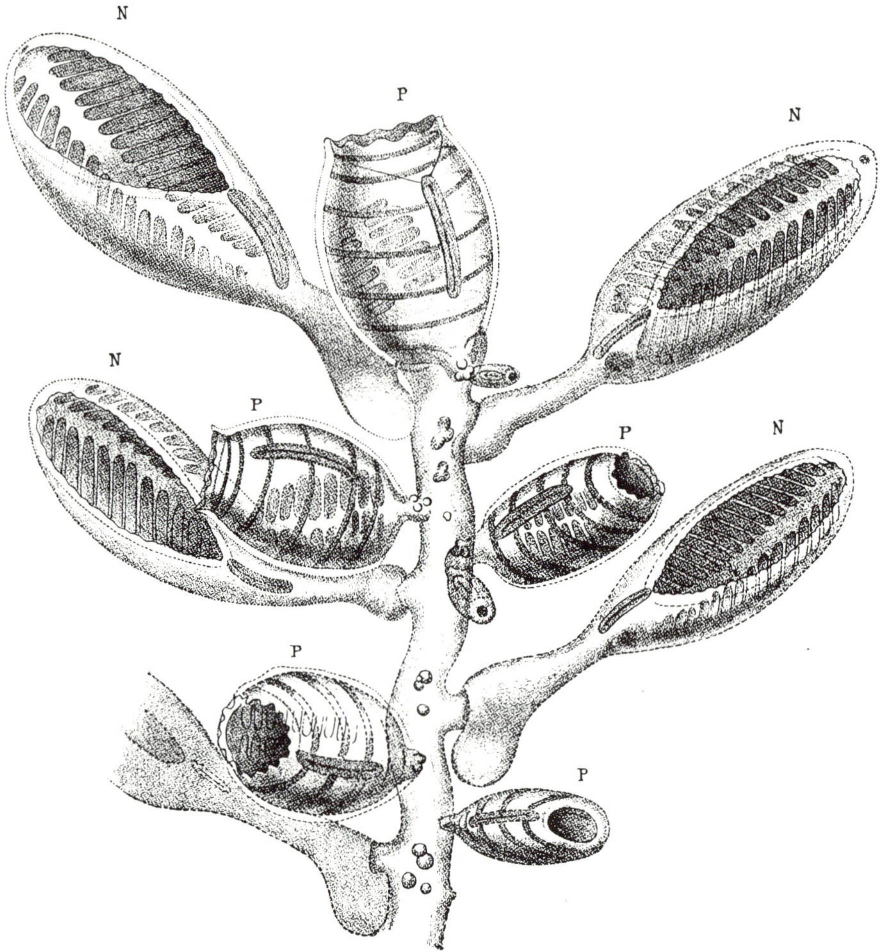

Figure 5.7. Part of a cadophore of the pelagic thaliacean *Doliolum* sp., showing gastrozooids (N) which perform feeding and respiratory functions, and phorozooids (P) which will detach as a subsequent, free-living stage. (From Bogert 1894.)

becomes a "deputy" that will eventually also differentiate into a phorozooid. The remaining bud of the cluster settles not on the oozooid, but on the stalk of the now-mature phorozooid. This bud gives rise to several new prebuds, which remain on the phorozooid stalk. The phorozooid then detaches from the mother oozooid. The "deputy" begins maturation into a phorozooid and the process repeats itself.

The detached phorozooid and accompanying buds, now free-living, continue the process. The buds on the phorozooid stalk are sexual pre-

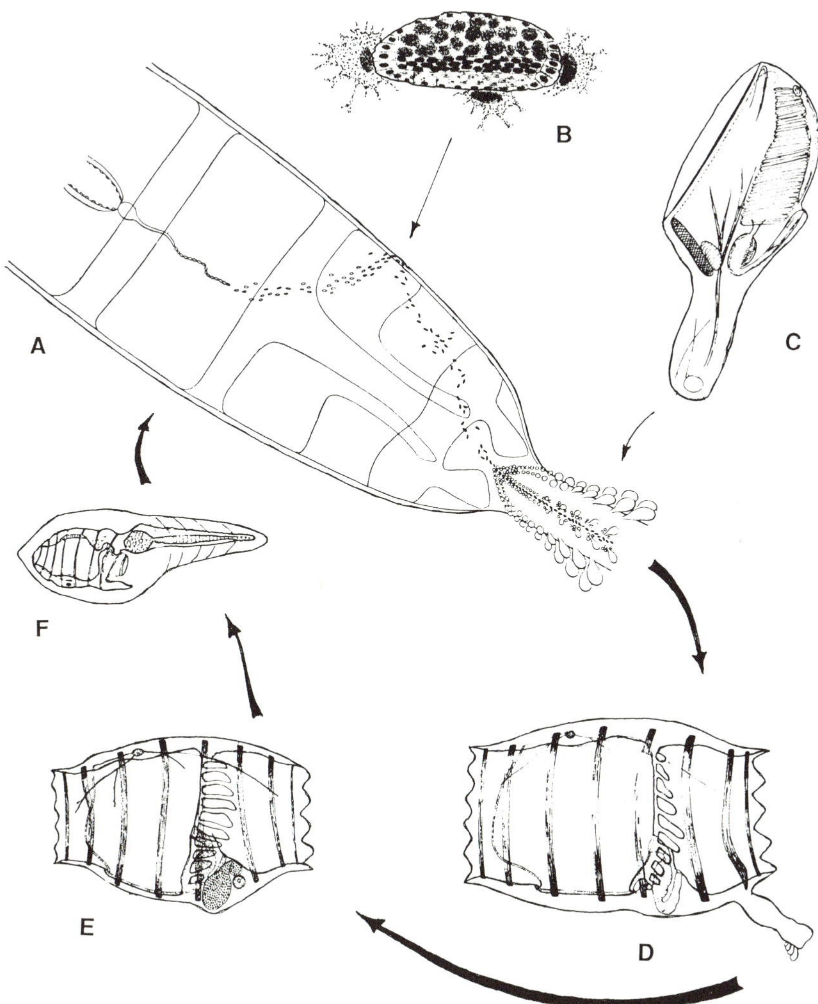

Figure 5.8. Generalized life cycle of *Doliolum*, a thaliacean. (A) Oozooid, the first postlarval zooid type; note migration of multicellular buds from ventral stolon to cadophore at lower right. (B) Enlargement of a bud being transported by three amoeboid phorocytes. (C) Gastrozooid and (D) phorozooid, which grow on the cadophore stalk of the oozoid. (E) Gonozooid, produced from buds on the stalk of a free-living phorozooid. (F) Larva, produced sexually by gonozooids. (A–E from Korschelt and Heider 1909; F from Herdman 1922.)

buds. These sexual prebuds grow and differentiate yet another zooid type, the gonozooid (figure 5.8e). Just as the mother oozooid bore gastrozooids and immature phorozooids in an earlier stage of the life cycle, now the phorozooid and gonozooids comprise another colony type. When gonozooids mature, they are sequentially released as yet another free-living stage, as solitary sexual individuals, which eventually release gametes.

Is this any way to build an animal? *Doliolum* and other invertebrates with complex life cycles are favorite exam questions precisely because instructors fail to provide a conceptual framework to account for this complexity. How might we explain the evolution of such a life cycle? We will attempt to provide an answer to this question in terms of different levels of organization at which natural selection has acted. But we will begin, incongruously but for reasons of comparison, with a digression into another better-known life cycle: the cellular slime mold *Dictyostelium*. This is a curious little creature that goes about its life in "disconnected bits and pieces." A spore gives rise to a free-living amoeba, which yields free-living asexual progeny. These aggregate, under adverse environmental conditions, to eventually form a differentiated fruiting body capped with spores.

The life cycle of *Dictyostelium* is useful for this discussion because it provides a forceful reminder that free-living cells, no less than organisms, have the capacity to reproduce, show heritable variation, and be selected. It is quite impossible to understand the life cycle of *Dictyostelium* without considering selection on both cells as free-living entities and organisms as aggregates of those same cells. In *Dictyostelium*, this point is particularly clear in the work of Filosa (1962), who showed that fruiting bodies are quite frequently composed of a mix of genotypes and that this mix typically includes forms that make little or no contribution to the formation of stalks (figure 5.9). Amoebae that will produce somatic stalk cells and others that do not may co-aggregate. Variants that do not produce stalks clearly gain whatever benefits there are from a stalk without apparently contributing to its costs. Hence, the life cycle of *Dictyostelium* is vulnerable to cells that fail to support somatic function (Buss 1982; Armstrong 1984; Matsuda and Harada 1990). *Dictyostelium* persists as a tension between selection on the individual level, favoring the differentiated state, and selection on the cellular state, favoring the loss of that differentiation.

This lesson may be applied directly to the *Doliolum* case, with one important modification. In *Doliolum*, there are four different types of

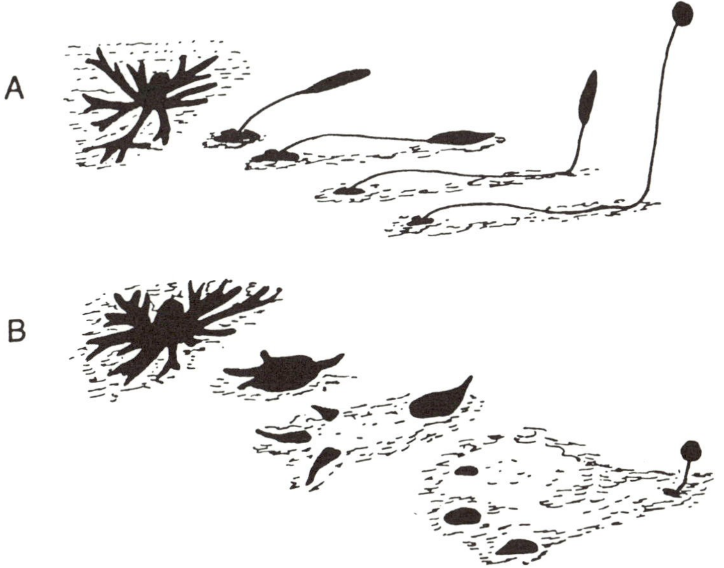

Figure 5.9. A "wild-type" *Dictyostelium* and a mutant that fails to produce somatic stalk cells. When the two forms coaggregate, the "wild-type" produces all somatic cells, to the relative benefit of the stalkless form. (From Bonner 1958.)

free-living individuals[1] and there are also zooids. Zooids are multicellular entities above the level of the cell and below the level of the individual; they are combined in different ways to form different types of individuals. They possess the capacity to reproduce, to display heritable variation, and, presumably, to be selected on the basis of that variation. If zooid-level variation is heritable,[2] then zooids are as discrete a unit-of-selection as are either cells or individuals.

[1] The term "individual" here is meant to designate a free-living, multicellular entity upon which selection acts as a unit. Each of the colony-types of *Doliolum* are treated as individuals.

[2] Zooid-level variation will fail to be heritable only if the germ line of *Doliolum* is fixed in early ontogeny, that is, if the original oozooid-derived prebud already has all gamete-producing cells terminally differentiated. There is no direct experimental evidence on this point; after all, *Doliolum*, as a fragile, open-oceanic form, is hardly a tractable laboratory system. Yet there are good reasons to believe that germ-line sequestration does not occur in this species. Consider the lineage of a medial prebud. After release from the oozooid, it divides to produce phorozooids and sexual prebuds. The latter then grow and differentiate a gonozooid, and only a subset of gonozooid cells gives rise to gametes. The original prebud, while multicellular, is minute. To retain a stock of determined cells sufficient to eventually dole out to all the sexual zooids it will eventually

In the *Dictyostelium* example, cells that abandoned supportive function may have been favored by virtue of their reproductive specialization. An active conflict exists between selection at the level of the cell and that of the individual (Buss 1982). While an extant conflict is not apparent in *Doliolum*, analogous conflicts, between selection at the level of the zooid and at the level of the individual, may well have shaped the evolution of this life cycle. The *Doliolum* oozooid is infertile, but one can easily imagine an ancestral form, as is the case in many colonial salps today, in which individual zooids played all roles. If a variant zooid type emerged that permitted the abandonment of supportive function, then that zooid type would be favored relative to those that retained supportive functions. Indeed, the existence of zooids specialized for reproduction is a common feature of many colonial organisms, most obviously occurring in bryozoans and hydrozoans. In *Doliolum*, however, the challenge is greater than that of *Dictyostelium*, bryozoans, or hydrozoans. Here one does not simply have sterile and generative zooids, but an entire sequence of sterile zooid types, and these zooids, along with sexual zooids, are combined variously into several distinct types of individuals.

There is little conceptual difficulty in dealing with the sequential appearance of different sterile zooids. Each sterile zooid has a different suite of functions: oozooids feed, respire, and asexually reproduce; gastrozooids feed and respire; phorozooids detach, feed, and respire. Since the initial prebuds have to be settled by specialized phorocytes on the oozooid, it is reasonable to assume that originally these were actually or potentially free-living units capable of all functions. A variant that retained that ability, as well as the ability to reproduce, may have generated the phorozooid. Finally, another variant that abandoned the somatic functions of the phorozooid could have given rise to the gonozooid. The process of the origin of zooid-level variation that favors one zooid type relative to another could conceivably give rise to this proliferation of zooid types.

The more difficult question is how these various zooid types could have come to be arranged in so many different types of free-living individuals. A naive observer who, say, collects an oozooid, an oozooid with associated gastrozooids and phorozooids, or a phorozooid with as-

---

produce would be quite impossible. While it is conceivable that prebuds maintain one or a few cells that are terminally committed for gamete production and that these cells simply become mitotically active after gonozooid formation, this habit, if it were to exist, would be a pronounced exception to an otherwise general pattern (Buss 1983).

sociated gonozooids would be confronted with three seemingly perfectly adequate and different organisms. However, as discussed above, each form is reached by passing through the other stages. It would appear that selection at the level of the individual is rather permissive; it has allowed the retention of multiple grades of organization.

We have, then, argued that life cycles of the *Doliolum* grade are produced when (1) zooids are units-of-selection; (2) zooid-level selection favors compartmentalization of function by sequential loss of somatic function in favor of proliferative potential; and (3) individual-level selection does not discriminate between the types of free-living individuals produced by zooid-level selection. What *Dictyostelium* and *Doliolum* share is not that selection must act on multiple units in these cases; it does so in all cases. Rather, these organisms simply make the matter particularly clear because their life cycles separate the units in a fashion that forces us to see as obvious a condition that may be easily obscured in the simple life cycles of unitary organisms.

There is a deeper lesson here as well. Recent treatments of the units-of-selection problem distinguish between replicators and interactors (Hull 1980; Dawkins 1983; Sober 1984; Lloyd 1989; Brandon 1990), replicators being a proxy term for genes and interactors, a term meant to include all those units upon which selection can act (e.g., chromosomes, plastids, organelles, cells, zooids, and individuals). This dichotomy usefully captures the intuition that we must contend not only with the unit of inheritance, but also with the multiple entities elaborated from directions encoded within the replicators.

This replicator-interactor dichotomy, however, is not a complete characterization of the scope of the problem. Replicators and interactors are combined and the specific form of their combination is what we call a life cycle. For example, the life cycle of humans alternates between a single-celled stage and a multicellular stage. Both the cellular and multicellular stages are units-of-selection. Consider RNA-phages. They cycle between a free-living (albeit in the only environment rich in endogenous replicases—a living cell) RNA molecule and an encapsulated dispersal stage. Both the free-living molecule and the encapsulated stage are units-of-selection. It is units-of-selection that alternate in life cycles.

The mapping of the replicator-interactor dichotomy onto life cycles remains a major unsolved problem in evolutionary biology. At least three routes of combining replicators and interactors into life cycles have been common in the history of life. The first is the physical enclosure of one unit within another in all stages of the life cycle, as occurred in the plasmids of plants, the mitochondria of all eukaryotes, and, perhaps, in the elaboration of the first cell. The second is the physical incorporation during only a specific stage in the life cycle, as is the case for many viruses, multicellular parasites, and symbionts. The third is the elaboration of one unit from the other to produce differing stages in a life cycle, as occurs in the development of a multicellular differentiated individual from a zygote. Combinations of these routes are what has generated the diversity of known life cycles. Yet, we still do not understand how interactors have been assembled into life cycles and what are, or have been, the limits on this assembly process.

We agree, then, with G. C. Williams (1975, p. 119) that "the main work of providing a workable theoretical superstructure for understanding the enormous diversity of life cycles remains to be done." This major problem lies ahead, and grappling with it, albeit in different terms, has been John Bonner's lifelong preoccupation. We suggest that only with the solution to this problem will today's neo-Darwinism complete its maturation into a genuine multiple-level theory.

Summary

John Bonner works that middle ground of biology, between molecules and organisms. He demanded throughout his career that the time would come when the day-to-day work of molecular biologists would shed light on the findings of evolutionists, and vice versa. That time is upon us. We offer three vignettes of research at the interface of development and evolution designed both to celebrate the quickening pace of progress and to identify problems that remain particularly recalcitrant. On the side of celebration, we note that the growing understanding of *Drosophila* early development is now sufficiently rich enough to permit century-old speculations in invertebrate zoology to be analyzed experimentally; we celebrate a newfound respectability in speculative zoology. From evolutionary theory, we argue that John Bonner's lifelong interest in levels of complexity is finding its way into the formal structure of evolutionary theory in the context of the units-of-selection debate, and argue that this will ultimately occur only by framing the issue as a problem in life-cycle evolution. Less worthy of celebration, but no less worthy of prominence, is the continued difficulty of understanding the source of major evolutionary changes in development. Here we argue that saturation mutagenesis, a technological tool of developmental geneticists, provides a little-appreciated but potentially important class of evolutionary data that bears directly on the issue.

Acknowledgments

We thank the organizers for the opportunity to participate in this celebration; N. Blackstone, D. Bridge, C. Cunningham, R. DeSalle, H. Horn, P. R. Grant, J. Reinitz, B. Schierwater, and E. Vrba for comments on the manuscript; and the National Science Foundation (BSR-8805961, OCE-9018396, DCB-9018003) and the Office of Naval Research (N00014-89-J-3046) for financial support.

6

Evolution of the Cell

MARC KIRSCHNER

During the past generation of frenetic activity in cell biology, most people have had little time to think about the evolution of the cell. Evolution on a macroscopic level has been the province of paleontologists. On a microscopic level, a great deal of interest has been shown in rates of mutation and phylogenetic distances derived from protein and nucleic acid sequence comparisons, but these studies usually have been pursued in a manner unconnected to cell biological or developmental mechanisms. Nevertheless, knowledge of protein sequences has led to more easily quantifiable evidence for common descent, as an unexpected level of conservatism has been found among enzymes and structural proteins in very distantly related species. Unfortunately, sequence information has so far provided little information about the ways cell biological processes have evolved.

In this review, I will concentrate on a discussion of the evolution of eukaryotic cells. Although a great deal of speculation has taken place on the relationship of prokaryotic cells to eukaryotic cells (Margulis 1981), it is rather unconnected to the fossil record and to the general body of information on evolution, which concerns the evolution of eukaryotic organisms. A discussion of the evolution of eukaryotic cells, although it does not address the fascinating questions of the origins of life, is nevertheless closely related to the origin of metazoan organisms and the evolution of developmental mechanisms. As I will show through some specific examples, recent progress in cell and molecular biology concludes that basic eukaryotic cell function has remained conserved over a billion years of evolution. Although this confirms the common descent of organisms, it offers little clue as to what has really evolved. To understand what has evolved in the eukaryotic cell we must consider the cell within the organism, or the cell as the organism. This means that the problem of cellular evolution cannot be considered separately from the problem of development itself.

Here I shall first consider the origin of new gene products, their contribution to evolution, and the various ways they have been used in cells of different organisms. Next, I will consider how cellular mechanisms using both old and new gene products have evolved and how these changes have allowed for organismal evolution to take place. We will see the particular value of cellular mechanisms that are easily adaptable and that can generate complex multiple outcomes, particularly processes such as transcription, cytoskeletal assembly, and signal transduction. Finally, I will consider the evolution of ensembles of cellular mechanisms, which in embryogenesis can be called developmental mechanisms. These give rise to fundamental embryological process such as gastrulation. In the end we shall find that the evolution of the eukaryotic cell can be most readily understood, not so much in terms of mutational change in the genes, but in the evolution of processes of development. Because we possess only a primitive knowledge of such processes, it may seem arrogant to speculate on their evolution, much less on the cellular basis for this evolution. However, there are greater sins than arrogance. The failure to think aloud about the origin of developmental mechanisms and their expression on a cellular level disallows us to speculate on the mechanisms of evolution, which biologists will never accept. Therefore, I will frame the questions and provide some ideas on the evolution of cells and development. My answers are meant more to show where we can look for appropriate answers in the next few years than to imply that, at this stage of our primitive understanding I have any of the answers.

Cells Have Evolved Very Little

An overriding conclusion from cell biological studies of the past decade has been the extraordinary conservatism of basic intracellular structures and processes. Although it has been known for many years that the major cell organelles are present in almost every eukaryotic cell (e.g., the nucleus, centrosome, Golgi, mitochondria) (Wilson 1925), the extent of this conservation on a biochemical level has come as a surprise to most biologists. The amazing similarity of proteins from diverse organisms has stimulated interest in the biology of organisms such as yeast, nematodes, fruit flies, slime molds, sea slugs, and frogs, all of which have experimental advantages over human beings. There is now a generally shared conviction that these organisms can be valuable not only for probing questions of evolution, but for providing answers to questions of human physiology and disease as well.

One example of conservation on a biochemical level has become clear recently: the regulation of the cell division cycle (Murray and Kirschner 1989a; Nurse 1990). It has been known for many years that whatever event initiates mitosis in different organisms, many of the same changes in cell structure occur as a result. In most organisms the nuclear membrane breaks down, the chromosomes condense, the interphase microtubule arrays dissolve, a mitotic spindle forms, the membrane systems such as the Golgi and endoplasmic reticulum disperse, other elements of the cytoskeleton may be remodeled, cell junctions and adhesions are weakened and so on. Cells from different organisms differ, but many of the general features are conserved over large phylogenetic distances. Until recently the process that regulates these disparate events was completely unknown. Recent studies from a variety of organisms have identified a 34kd protein kinase as the principal enzyme responsible for initiating the transition from interphase to mitosis. The kinase acts directly on some substrates, such as the nuclear lamina, and indirectly on others, such as S6 kinase, to cause disassembly of the nuclear envelope, and is presumably responsible for the other events of mitosis as well (Ward and Kirschner 1990; Peter et al. 1990). Inactivation of this kinase at anaphase sets in motion the reversal of these processes, where all of the mitotic phosphorylation events are undone by constitutively active phosphatases.

The p34 kinase is activated by the accumulation of a set of unstable proteins called cyclin proteins. When these proteins reach a threshold, the 34kd kinase is activated and the cell enters mitosis. The cell exits from mitosis and returns to interphase when the cyclin molecules are abruptly degraded at the metaphase-anaphase transition. Cyclin proteins were first identified in sea urchin eggs but are now thought to be present in all eukaryotic cells (Evans et al. 1983; Pines and Hunt 1989). Although the 34kd kinase was first identified in the fission yeast, *Schizosachromyces pombe*, a homolog of this protein has been found in all eukaryotic cells. This homolog shows strong sequence similarity in all eukaryotes but, more amazingly, it shows functional similarity as well. The human p34 gene, when placed into yeast, will satisfactorily replace the natural yeast gene for all of its known functions, which include the ability to mediate the G1/S as well as the G2/M transition (Lee and Nurse 1987). Other major proteins regulating the p34 kinase such as the cdc25 activator and the weel inhibitor are also highly conserved and generally functionally interchangeable between organisms as distantly related as yeast, sea urchins, and human beings (Sadhu et al. 1990; Booher and Kirschner, unpublished).

The argument for conservation is strongest when functional tests, such as gene replacement or in vitro complementation, can be applied. Most often, though, we must rely on sequence comparisons. But do these sequence comparisons monitor adaptive evolution, or do they monitor genetic drift? In the case of p34 there is extensive sequence identity throughout the molecule between yeast and humans, which diverged more than a billion years ago. In the case of the conserved regulators of p34, cyclin, cdc25, and weel, the identity is limited to a small portion of the molecules and there is extensive divergence in other domains. Yet in the case of cdc25, for example, despite the large sequence divergence, the human molecule will complement yeast mutants (Sadhu et al. 1990) and in the case of weel, frog can also complement yeast (Booher and Kirschner, unpublished). In an in vitro system, sea urchin cyclin, despite its large divergence in most of the molecule from the frog cyclin, will complement a deficiency of frog cyclin. These results suggest that functional divergence of these important regulating genes has been minimal, while sequence divergence has been extensive. Since cdc25, weel, and the p34 kinase fully complement deletions of these genes in species that diverged more than a billion years ago, we can conclude that no important yeast functions are missing in the human protein. The reciprocal experiment is not possible in humans but may soon be possible in mouse (Thomas and Capecchi 1990). Therefore, while these sequences have apparently drifted extensively, they do not appear to have evolved functionally very much.

With the help of genetic tests, the list of highly conserved cellular functions has continued to grow. In some cases phylogenetic barriers have emerged, but for many systems they are minor and easily overcome. For example, the β-adrenergic receptor that normally responds to catecholamines in heart muscle will not function in yeast to replace a related receptor that responds to mating pheromones. However, addition of one more element to the signaling system, the mammalian Gα protein, will allow the yeast cell to respond to catecholamines and undergo the mating response (King et al. 1991). The obvious conservation of DNA structure has been matched by the conservation of histones, transcription factors, splicing enzymes, ribonucleoprotein complexes, and nuclear pores. The well-known conservation of the protein synthesis machinery has been extended to protein secretion, including components of the endoplasmic reticulum and Golgi. The major cytoskeletal proteins such as tubulin and actin have been conserved and serve similar functions in many organisms, as do the metabolic enzymes. The signaling and regulating molecules such as ras and hormone receptors, the regulating kinases such as protein kinase A, protein kinase C, S6 kinase—all have been found in every eukaryotic cell. There have been, of course, new inventions, expansion and specialization of each repertoire, but, as we shall see, many of the new demands of specialized cell types have come about by usurpation of existing components. Viewed as a computer, we would have to say that the basic hardware is similar in all eukaryotic cells; if anything has changed, it is the software.

Software Changes in the Cell Cycle

Though the proteins that regulate the cell cycle may be nearly identical in all organisms, the strategy for regulating the cell cycle is not. In the frog egg the accumulation of cyclin to a threshold initiates mitosis, and this process is independent of any transcriptional control (Murray and Kirschner 1989b). In the *Drosophila* embryo after cellularization, cyclin accumulation is also required for mitosis, but it is not the regulator. In this case control of the mitotic process is under control of the mitotic activator cdc25, and its expression is under transcriptional control (Edgar and O'Farrell 1990). Recent studies of cyclins specific to the G1/S transition in yeast have shown that their accumulation is under transcriptional control, but also may be under posttranslational control (Wittenberg et al. 1990; I. Herskowitz, pers. comm.). In cleaving frog eggs and sea urchin eggs, cyclin accumulation is completely unregulated. However, in the case of the G1/S cyclin in yeast, the accumulation is tied to the whole pathway of the mating pheromone response as well as to other less well understood pathways involving cell size and nutrition.

In the few well-studied cases of cell cycle control we can see both conservation and divergence. The major components are highly conserved and most are functionally interchangeable. The basic reaction pathway involving cyclins, p34 kinase, and other kinases and phosphatases is also identical. Yet the rate limiting steps and their linkage to other processes are different in different cells. This has enabled the cell cycle control mechanisms to respond transcriptionally to spatial signals, to be linked to extracellular cues, to be coupled to various homeostatic mechanisms, or to operate nearly autonomously during the rapid cleavages in the early embryo. As with modern computers, the architecture of the machine allows for many software applications. One might even say that computer hardware has evolved to allow for greater software flexibility. As we shall see, much the same can be said for cellular mechanisms.

Divergent Pathways of Photoreception

The cephalopod eye and the vertebrate eye are exquisite examples of convergent evolution. The anatomy suggests that the origins are totally independent. The vertebrate eye develops as an outgrowth of the brain; the cephalopod and insect eye develops as a peripheral ectodermal structure that grows into the brain (Young 1974). The topology of the nerves and photoreceptors is reversed. In the vertebrate eye, light passes through the nerves to the photoreceptor; in the cephalopod eye, light impinges directly on the photoreceptors. Is this anatomical convergence reflected in a totally separate origin of the biochemistry of photoreception?

The key event in photoreception, the photoisomerization of retinaldehyde, has been widely used. In prokaryotes, where it is part of the proton pump, and in eukaryotes, where it is used as a photoreceptor, retinaldehyde, which is chemically the same in all systems, is bound to an integral membrane protein called opsin, whose polypeptide chain spans the plasma membrane seven times. There is no sequence homology between the prokaryotic opsins and the eukaryotic opsins, though overall structural similarities in the positions of the amino and carboxyl ends and the number of transmembrane helices suggest that at one time these proteins could have had a common origin (Henderson and Schertler 1990).

In eukaryotes, whether cephalopods or mammals, opsin is a 7-membrane spanning protein, and all such proteins are receptors that are thought to couple to intracellular GTP binding proteins called G-proteins. This widespread family of membrane protein receptors includes the receptors for the mating pheromones in yeast, the cAMP receptor in slime molds, and the serotonin and β-adrenergic receptor in mammals (King et al. 1991). The receptors catalyze the exchange of GTP for GDP on the heterotrimeric G protein. Binding of GTP causes dissociation of the trimeric G protein into Gα and GβGγ; these subunits interact with other cellular enzymes and regulate their functions. The invertebrate opsins, which are 7-membrane spanning integral membrane proteins, have clear sequence similarity to the vertebrate opsins (Yokoyama and Yokoyama 1989). In the central region of the molecule there is also a very strong similarity on the nucleic acid level, and throughout the molecule there is extensive similarity with a few insertions or deletions. There is no question that rhodopsin, the primary unit of photoreception, has evolved from a common precursor.

The vertebrate opsins are known to couple to a heterotrimeric G protein called transducin, which in its GTP form activates directly a cGMP phosphodiesterase. In the vertebrate photoreceptor, increased levels of cGMP open a Na^+ channel leading to increased neurotransmitter release. Therefore, the action of light causes a drop in cGMP and an inhibition of transmitter release that inactivates an inhibitory neuron, which ultimately leads to elevated electrical activity in the brain (Stryer 1988). The vertebrate photoreception system also has a means of adaption that desensitizes the receptor after stimulation. It involves the binding of a small protein, called β-arrestin, to the cytoplasmic domain of the receptor after a period of activation (Bennett and Sitaramayya 1988).

In invertebrates, although the initial coupling of opsin to signal transmission are similar, the complete pathway is designed differently. *Drosophila* is known to contain G-proteins (Guillen et al. 1990), and the structure of invertebrate opsin strongly suggests that the receptor couples to G-proteins; the exact G-protein that couples to *Drosophila* rhodopsin is not known. Like vertebrates, *Drosophila* contains a β-arrestin molecule that is highly conserved, suggesting that *Drosophila* rhodopsin contains the same desensitization system as mammals (Smith et al. 1990). However, the next part of the pathway seems divergent. G proteins are known to couple to several second messenger systems, and the best evidence suggests that G protein in invertebrates (*Drosophila* and the horseshoe crab, *Limulus*) couples to a different second messenger system from that affecting cGMP phosphodiesterase (Suss et al. 1989; J. Brown, pers. comm.). Genetic approaches can be useful in delineating this second messenger pathway. Recently, *Drosophila* mutants have been obtained that have morphologically normal cells that do not respond to light. The gene that is defective in one of these mutants has been cloned and shown to have strong similarity to phospholipase C, an enzyme involved in cell signaling (Bloomquist et al. 1988). There is evidence that Ca^{++} release, mediated by inositol triphosphate, occurs during light stimulation, which suggests that in the invertebrate photoreceptors the G protein linked to opsins may activate phospholipase C and signal either Ca^{++} pathways via inositol triphosphate or protein kinase C via diacylglycerol. It is also possible that the unknown G protein signals some other second messenger pathway. Downstream of this signaling system there is an increase (as opposed to the decrease in vertebrate photoreceptors) in a nonselective cation channel leading to a *depolarization* and secretion. Thus the invertebrate system uses the same visual pigment, an evolutionarily related receptor, a very similar desensitization system; but most likely it couples this receptor to a different G-protein-mediated system to produce the opposite electrophysiological result from the one that occurs in vertebrates. In the end the brain still gets the signal.

The lessons of the comparative physiology of vertebrate and invertebrate photoreceptors is that the basic components have been highly conserved but their linkage has developed differently. The basic input of photons is the same; the output hyperpolarization or depolarization of the photoreceptor cell is completely different. In between there has been a high degree of conservation: retinaldehyde, 7-membrane spanning receptors, G proteins, β-arrestin, phospholipase C, nonselective cation channels; but the circuitry is different. The evolutionary invention was not in the types of proteins but in software for linking signaling and responding pathways together.

New Components and Their Evolutionary Value

Not all the remodeling of the eukaryotic cell is the equivalent of rearranging the furniture. There are, of course, new genes whose expression facilitated rapid evolutionary change. In the computer analogy these are the hardware improvements, which often provide new capacities for software innovations. As we shall see, some of these new genes may have persisted underutilized for extended periods of time, until the appropriate software mechanisms were developed to make use of them. In most cases the origins of these genes is traceable to more primitive structures that were stitched together by gene duplication and exon shuffling, but in some cases there is little clue as to their origins. It seems likely that some of these specific genes are crucial for major branches of macroevolution. Although one can tabulate many genes that would qualify as a "great moment in evolution," I will discuss only two structures dependent on new genes that are important for the major radiations within the vertebrates: myelin and feathers.

The biophysical features of nerve conduction explained by cable theory show that the rate and efficiency of nerve conduction increase with the diameter of the nerve fiber and with the decrease in the capacitance of the plasma membrane. To process complex information or to respond quickly to a predator or to capture food, rapid nerve conduction is obviously advantageous. In invertebrates, conduction velocity and efficiency is increased by increasing the diameter of the axon. In cephalopods, such as squid, it has reached its zenith in the giant axon—0.5 mm in diameter. To construct a large central nervous system with giant axons is a difficult packaging problem.

In mammals, the problem of conduction velocity and size of the neuron has been solved principally by myelin, an electrical insulator that decreases the capacitance of the membrane and causes the action potential to move by saltation from one node or gap in the myelin to the next, about every millimeter. Myelination requires the wrapping of compressed cell membranes tightly around the nerve. These membranes are provided by oligodendrocytes in the central nervous system and Schwann cells in the peripheral nervous system. The tightly wound plasma membranes must exclude cytoplasmic contents and be able to be brought in close apposition on both their external and internal faces. Both types of cells contain an 18kd protein called "myelin basic protein," which is thought to aid in the collapse of the cell membranes by excluding cell contents, as well as specific transmembranes components such as the Schwann cell glycoprotein PO, and the oligodendrocyte proteolipid, which are thought to help bring the extracellular domains in apposition. The means by which the PO glycoprotein brings the extracellular domains together is evident from the structure of the protein (Lemke et al. 1988). The PO protein shows the structural motif of a single immunoglobulin domain on the extracellular surface. This is a very ancient motif and is shared with a number of proteins of quite distinct functions, such as immunoglobulins, neural cell adhesion molecule (N-CAM), PDGF receptor, the Fc receptor, and protein on the surface of T cells such as CD4 and CD8 (Williams and Barclay 1988). This motif is well defined structurally, but what do these membrane proteins have in common? It is thought that they arose from early cell surface proteins that are involved in homotypic cell surface interactions, and that these early homotypic interactions evolved later into heterotypic interactions such as the Fc receptor, immunoglobin molecules, or growth factor receptors. The PO protein, constituting 50% of the protein of peripheral myelin presumably, uses these primitive homotypic interactions between the outer surfaces of the plasma membrane to promote its collapse, causing the folded cell membrane to stick to itself. Another protein in peripheral myelin, the P2 protein, shows its origin quite clearly. It has sequence homology to a class of lipid-binding proteins, such as the fatty acid binding protein of the liver or the retinal binding protein (Lowe et al. 1985), and may act to allow internal opposition of plasma membranes and other vesicular components.

Invertebrates do not seem to have invented myelin basic protein, glycoprotein, PO, or myelin proteolipid, and therefore the invention of myelin probably deserves to be considered as a great moment in vertebrate evolution. This compact insulator allows the generation of an efficient, fast, and compact nervous system. Though devoid of myelin, invertebrates have made a feeble attempt at insulation, using the sheath membrane from interstitial cells, such as those found in crustacean nerves (Kishimoto 1986). Since this membrane does not wrap several times around axons, it has limited insulation capacity and a limited improvement in conduction velocity. Interestingly, although invertebrates have not invented myelin, their sheath membrane contains characteristics of sphingolipids like those found in myelin membrane. The type of sphingolipid in the sheath membrane of invertebrates can be used to divide the chordates and echinoderms (so-called dueterostomes) from the other invertebrates (so-called protostomes). Thus the primitive chordates contained interstitial cells with the proper sphingolipid for making myelin, and yet the nervous system of tunicates and amphioxus hardly seems advanced when compared to invertebrate protostomes like insects and molluscs. It is possible that other aspects of myelin such as myelin basic protein or PO arose in primitive chordates along with the sphingolipid. Certainly IgG binding domains are known in invertebrates as well as vertebrates (Bieber et al. 1989), where they serve as homophyllic binding domains. If all of this is true, the structural ingredients for myelin may have existed for a long time, yet the realization of its value certainly did not come with sea urchins, tunicates, and other simple deuterostomes. The true "invention" of myelin occurred with the vertebrates. Like unused computing capacity, the invention of myelin was permitted by preexisting ingredients but realized only with the generation of developmental processes for linking these gene products together in a new cellular program, which in turn may have been dependent on the new selective advantage of developing a complex nervous system that occurred in the early evolution of the vertebrates.

Evolution of Scales and Feathers and Intermediate Filaments: New Genes and Old Genes for New Purposes

By mass alone, intermediate filaments are the most important intracellular structural proteins. They have two important properties: exceptional tensile strength and the capacity for dynamic assembly and disassembly. In vertebrates, their evolution seems closely tied to the evolution of tissues. For example, intermediate filaments make up the total structure of hair, most of the structure of the cornified epidermis, and provide most of the tensile strength for all epithelia; all of these structures are composed of a subfamily of nineteen different keratin-type intermediate filament proteins. All of the keratin intermediate filaments are heterodimers of two general classes of keratin filament proteins. Neurofilaments and peripherins are intermediate filaments that are prominent components of nerve axons. Glial filaments form intracellular structures, in glial cell processes; vimentin filaments are structural components of all mesenchymal cells; desmin filaments are structural elements of muscle; and nuclear lamins make up the major structural component of the nuclear envelope. The entire family of intermediate filaments, but particularly keratins, has been well studied by biophysical techniques. They are all coiled-coil helial dimers with large C-terminal tails and they assemble into 10 nm filaments. The sequences of the large number of intermediate filament proteins are known in many organisms so that the sequence evolution is comparatively easy to trace (Stewart 1990).

Until recently all the molecular information about intermediate filaments came from vertebrates. Recently, sequence information of invertebrate intermediate filaments has begun to clarify the evolutionary picture. The ancestral intermediate filament seems to be the nuclear lamin, found in all metazoan cells but not yet identified in yeast. The overall structure of lamin is closely related to other intermediate filaments, with the sequences at the end and beginning of the coiled-coil domains showing close sequence identity. One structural feature distinguishes lamins from other intermediate filaments: the coiled-coil domain in lamins is larger than that of any other class. It seems likely that lamins, which are present in virtually all cell nuclei, possessed this extra region of coiled-coil helix, which was lost in all the other vertebrate intermediate filaments.

Invertebrates seem to have a more limited diversity of intermediate filaments than vertebrates. It now appears that arthropods do not have any intermediate filament other than lamins and therefore this lineage does not seem to have "discovered" their potential for intracellular cell structure. Recently the intermediate filament of snails and nematode epithelia have been sequenced and analyzed by Klaus Weber and colleagues (Weber, Plessman, et al. 1988; Weber, Plessman, and Ulrich 1989). Despite their presence in epithelia, these filaments are not keratinlike. They share the more ancient expanded coiled-coil domain of nuclear lamins. This sequence conservation extends beyond the coiled-coil domain into the usually highly divergent C-terminal domain region, and therefore these cytoplasmic epithelial intermediate filaments would be classified as a member of the nuclear lamin family. Though there may be more intermediate filaments in invertebrates, they do not seem to be of the typical vertebrate types and may have also derived directly from the nuclear lamins (H. Gainer, pers. comm.). None of the invertebrate intermediate filaments found in epithelia so far are like vertebrate keratins: they show neither the typical keratin structure or the heterogeneity of vertebrate keratins, nor are they heterodimers, as are the vertebrate keratins. In vertebrates there is a rigorous separation of mesenchymal (vimentin) filaments from epithelial (keratin) filaments, yet this does not seem to be the case in the snail, where the same classes are found in several tissues. Weber thinks that the evolution of the complex family of keratin filaments in vertebrates parallels the evolution of complex epithelia, such as stratified and transitional epithelia. He speculates that the evolution of these cell types may have depended on the invention of keratin out of the ancient family of laminlike intermediate filaments.

In tracing the origin of intermediate filaments, we can conclude that the basic invention, nuclear lamins, occurred very early in eukaryotic cell evolution but that the extraordinary diversity of intermediate filaments only occurred in the vertebrate, or perhaps the chordate, radiation. Of course, we do not know whether some precursors to keratin or other intermediate filaments existed in animals that disappeared in one of the great extinctions, but there is as yet no trace before vertebrates. Intermediate filaments are another example where the basic structural invention was made early in evolution, where it was widely utilized for one purpose, generating the structure of the nucleus, where this was extended in some animals to a limited degree and in others (arthropods) not at all, but where it was elaborated and diversified by a single group of organisms to great effect.

The evolution of the integument of animals is very important in their ability to occupy new environments. Scales in reptiles and birds and feathers in birds represent extraordinarily important innovations in the integument and essentially define these classes. There is strong sequence information that suggests that the major protein of feathers arose from the major protein of scales by several small insertions and deletions in the scale gene. The origin of the original scale gene, however, is a complete mystery. Despite both being called β-keratins, the scale and feather proteins are totally different structurally and on a sequence level from the α-keratins of stems, hair, and so on, and from the collagenous and often mineralized scales of fish (Sire 1989). The β-keratin protein of scales and feathers is a 10kd β-pleated sheet structure, while the α-keratins are 50–70kd proteins that have an α-helical coiled-coil domain.

Birds contain a large number of genes coding for β-keratin, which is used for claws, beaks, feathers, and so on with about twenty coordinately expressed during feather growth alone (Presland et al. 1989). In chickens these genes are tandemly located in the genome, and each shows an extraordinary degree of conservation in the coding sequence (Presland et al. 1989). Among four adjacent genes sequenced, there is 94% identity on the DNA level. To maintain this high degree of sequence similarity over 140 million years of evolution requires an active process such as gene conversion. Rogers and his colleagues have suggested that it is necessary to control the diversity of individual feather genes because a single mutation in a structural protein of a complex structure like the feather could disrupt the oligomeric assembly of other unmutated subunits. The deleterious effects of such mutations of oligomeric structures has recently been demonstrated by the identification of several dominant mutations in human collagen genes that cause severe skeletal malformations such as osteogenesis imperfecta (Byers 1989). Therefore, once elaborated by mutation of scale genes and by subsequent duplication, the feather genes of birds must have been maintained in a state that could not easily be altered.

The scale and feather keratins co-evolved with reptiles and birds. There is no evidence that these protein motifs lay around for millions of years, like the intermediate filaments. However, once invented or rediscovered, the birds duplicated the original scale genes, retained some for scales, while the rest evolved into more specialized β-keratin. These feather genes were actively maintained in an amplified form as homogeneous sequences by gene conversion, in which state they supported high rates of synthesis. The impact of these genes was immense. They allowed birds and reptiles to escape the restrictions of a watery environment, as well as freeing birds from the purely terrestrial environment. It is interesting that vertebrates never made use of chitin, a structure that enabled invertebrates to escape the aqueous environment. Thus the origins of scales and feathers are truly great moments in evolution bestowed only on these two classes of vertebrates where the origin of these unique genes remain totally obscure.

Evaluating the Important Evolutionary Changes in the Genome

Paleontologists and molecular biologists have the same handicap in looking at the fossil or nucleotide record; they are looking at successful solutions. They have extreme difficulty in identifying the selective pressures on these organisms and the range of possibilities from which the successes were selected. They cannot with certainty distinguish important changes from unimportant changes. For molecular biologists, it is even bleaker: all amino acid changes look the same, and it is difficult even to speculate about their fitness. To speculate intelligently on the course of evolution it is necessary to understand what genetic changes made major differences in evolution. As I argued, all of the important changes seem to have occurred despite the conservation of many structural and regulatory genes and with only a limited, but at times, crucial input of novel gene products. It seems that many of the important changes in evolution must have been in the strategies that recombine these fundamental processes for new purposes. Even truly new inventions such as intermediate filaments may have lain dormant at first, conferring small or specialized advantages until they were put to new uses by new developmental mechanisms. Therefore, to identify the important changes in cell function that allowed major new evolutionary directions, we have to consider the evolution not of specific gene products but of cellular pathways. These pathways made use of other gene products that we have not considered—regulatory elements in the cytoplasm and nucleus—whose own evolutionary change plays a large role in the evolution of metazoan organisms.

Our analysis must begin with development, during which the embryo increases dramatically in complexity. Modern work suggests that very little information is prelocalized in the egg. What the egg lacks in complexity of organization it possesses in the complexity of instructions in its genome and in its cytoplasmic organization, which together are capable of generating many cells and many different types of cells in complex patterns. This capacity for generating complexity out of the initial limited complexity of a single cell is also found in some somatic cells, such as the stem cells of the blood cell lineage. In evolution cells must have acquired the capacity to generate variability both in gene expression and in cell behavior. The mechanisms for generating complex cell types with complex morphologies and then organizing them in new tissue arrangements are the important processes that are constantly changing in evolution. These processes may be broadly called "regulatory" processes because they link several biochemical mechanisms together.

I shall examine three regulatory processes: transcriptional regulators and the control of gene expression in the nucleus; microtubule assembly and the regulation of cell morphology in cytoplasm; and G-proteins and the organization of extracellular signaling systems in the plasma membrane. I shall examine these three processes to see how susceptible they are to genetic change. All of these systems have unusual properties that allow a small number of gene products to be used in a large number of developmental situations. These modular properties also make them sensitive to mutation with pleiotropic effects and allow rapid evolutionary change. We shall see that the very properties that make these systems adaptable to a variety of environmental circumstances also seem to have the ability to generate a high proportion of functional outcomes in evolution. In the end it is not these mechanisms themselves, but ensembles of them which are the appropriate level for understanding evolutionary change. In the case of embryonic change, I shall call them "developmental mechanisms."

Generation of Complexity in Transcriptional Control

All organisms regulate their protein content in response to environmental and developmental signals. The elegant work of Jacob and Monod set out a paradigm for transcriptional control that has accurately described the regulation of many bacterial and some eukaryotic transcription units. In this model specific transcriptional activators or repressors bind to sequences on the DNA and positively or negatively regulate the activity of RNA polymerase, which binds to a DNA sequence called the promoter. Britten and Davidson extended this model of regulation to a hierarchy of regulatory circuits that could control batteries of genes. The existence of some sort of regulatory circuit is undisputed, and recent evidence from studies of early embryogenesis of *Drosophila* suggest that the specification of tissues can be regulated by cascades of transcription regulators (Biggin and Tjian 1989). Recently, a great deal of new information has become available about transcriptional regulation in eukaryotic cells. We can now look at the nature of the transcriptional circuitry to try to understand how it contributed to the evolution of cell and organismal complexity.

Although some elements of the Jacob-Monod repressor scheme for bacteria and their viruses exist in all organisms, the majority of regulation in eukaryotes looks superficially very different. In bacteriophage lambda, the alteration in spacing of a few nucleotides inactivates transcriptional activation (Ptashne 1988), whereas in many eukaryotic "enhancers" the sequences for binding specific proteins can be moved over long distances, their orientation can be reversed, or they can be present upstream or downstream of the promoter site for the initiation of transcription. The enhancer sequences bind a myriad of factors that interact in positive and negative ways indirectly with the eukaryotic polymerase, which itself is a much more complex molecule than the prokaryotic polymerase (Alberts et al. 1989; Darnell et al. 1986).

Why has the eukaryotic cell employed such complex and apparently imprecise mechanisms for transcriptional control? Why does it use proteins that operate at a distance, and how can it construct transcription complexes with so many degrees of structural freedom while the typical prokaryotic complexes are so severely constrained? At first, the puzzling behavior of enhancers seemed to suggest that the answer lay in the fundamental difference between the eukaryotic and prokaryotic cells, namely, the presence of a nucleus. It is unlikely however that the difference between the prokaryotic repressor and eukaryotic enhancer is due to different nuclear structure. Prokaryotic cells are now known to have enhancer sequences (Wedel et al. 1990), and the function of eukaryotic enhancers seems not to involve the intricacies of chromatin or nuclear structure. Instead, transcriptional regulation in eukaryotes involves many of the same principles of sequence-specific DNA binding interactions as does the lambda repressor. Unlike the short-range factors prominent in prokaryotes, these transcription factors require strong protein-protein interactions that allow complexes between several DNA binding proteins to be assembled over long distances. The short-range interactions in most prokaryotic transcription complexes make use of specific binding energy and geometry provided by the DNA to assemble the complexes; enhancers cannot.

There are obvious disadvantages to long-range elements: they must involve stronger protein-protein interactions to compensate for distance, and they are in danger of being interfered with or removed by unequal crossing over or insertional mutagenesis. The advantages possessed by long-range regulators seem to be their versatility and their sensitivity to mutation. It is now clear that the same transcriptional elements can be used as both positive and negative regulators in different contexts and can act synergistically to control transcription (Lin et al.

1990; Diamond et al. 1990). This combinational means of generating diversity of transcriptional control avoids the complicated progression of transcriptional regulators implicit in the original Britten and Davidson cascade model. The detailed properties of these long-range interactions are just being worked out, but one unusual feature that has emerged is the relative nonspecificity in protein-protein contacts. The first indication of this came from studies on the Gal4 enhancer in yeast (Ptashne 1988). Ptashne showed that transcriptional activation in this typical eukaryotic enhancer consisted of a well-defined DNA sequence binding element attached to an activation domain. The activation domain had remarkably loose structural requirements; virtually any acidic sequence attached to the DNA binding element would activate transcription. The loose sequence requirements for the enhancer activation domain contrasted with the very strict structural requirements for short-range transcriptional activators such as the lac repressor. It may be that a typical eukaryotic gene may contain many different upstream control sequences that vie or reinforce each other to produce a net positive or negative response. The three-dimensional structure of these complexes is not known, but it is likely that they are sloppy aggregations of a number of reinforcing or inhibiting contacts with the core transcriptional machinery.

The advantage in these sloppy arrays of DNA binding factors as opposed to the highly refined structure of the repressor is the flexibility it affords. Transcription complexes can be added to, rather than be completely redesigned. These complexes have not been further specified in the cell, perhaps because there is an advantage to modifiability.

Having regulatory sites that are dissociated from the promoter, the position of transcriptional initiation produces a system that is sensitive to insertional or recombinational mutations. The sensitivity of the transcriptional machinery to mutation may have allowed for rapid remodeling of controls during evolution (Yamamoto 1989). Enhancer elements can be moved around easily, which will put genes under new influences. It is now clear that some enhancers are present on movable genetic elements, greatly facilitating reorganization of transcriptional controls. Some of these movable elements may have acquired oncogenes and are now recognizable as tumor viruses (Bishop 1987). Others, such as avian leukosis virus, activate cellular protooncogenes such as c-myc by inserting near them in the genome (Payne et al. 1981). Since the work of Barbara McClintock, it has been known that movable genetic elements may be responsible for the activation of specific genes under a developmental program of maize (see, for example, Federoff 1989).

Most recently, the capacity of a transposable element to confer specific regulations of a cellular gene has been seen clearly in yeast. Yeast cells contain about twenty Ty elements, which have diverged somewhat from each other; some encode functional transposases. Insertion of the Ty element can activate the transcription of cellular genes, such as one isozyme of cytochrome C, which is essential when the other isozyme is inactivated (Errede et al. 1987). Cytochrome C then comes under developmental control because the Ty transcriptional enhancer sequence is inhibited in the diploid form of yeast by a specific DNA binding protein expressed in diploid cells, called a1-α2. The Ty element can also act as a transcriptional inhibitor (Roeder and Fink 1982), and recombination events can put this under developmental control. Another transposable element in *Drosophila*, copia, carries the heat shock element (Strand and McDonald 1985). In other cases there is recent evidence that transposable elements bearing enhancer sequences have brought new genes under regulation. The mouse sex-limited protein is 95% similar to the neighboring T-lymphocyte cell surface protein, CD4. Somehow, in mouse but not other rodents, an androgen-dependent enhancer has inserted itself close to the gene and brought it under androgen regulation (Stavenhagen and Robins 1988). There is no apparent value to this particular regulation.

Why would organisms retain potentially dangerous movable genetic elements? This mutability represents a burden, albeit a small burden for the individual. However, in the past such genetic elements may have facilitated evolution. Such a fundamental mechanism important for earlier radiation may be very difficult for the individual to dispose of. If the burden is small for the individual and historically deeply rooted, it will probably be maintained. Mating-type transposition in yeast may be a residue of an earlier time when transposition was not only a mutational event with evolutionary advantages but an adaptive one as well (Herskowitz 1989). The use of gene arrangements for rapid adaptation has been discussed by Borst and Greaves (1987) in the case of antigenic variation in trypanosomes. In these protozoan parasites the surface glycoprotein gene casette is transposed from an inactive site into an active transcription site, resulting in the expression of new cell-surface proteins. The trypanosome is preadapted to a new environment. Such mechanisms can be thought of as adaptive mechanisms for instant evolutionary change; they may have been prominent in primitive organisms and retained as specializations in a few cases.

Mechanisms for Generating Complexity in the Cytoplasm

Just as specific transcriptional mechanisms for generating patterns of gene expression have evolved, so cytoplasmic mechanisms have developed for generating complexity of cell shape. Cells adopt different morphologies in their developmental experience. There is great resilience in these morphogenetic processes. Neural crest cells can be transplanted into foreign pathways and migrate normally and differentiate appropriately (Newgreen and Erickson 1986). Optic nerves can migrate and connect to the proper locations in the optic tectum (Nakamura and O'Leary 1989). Cells can enter mitosis from different morphologies and generate functional mitotic spindles. They can recover from artificial perturbations, such as transient depolymerization of the microtubules or mechanical manipulations of chromosomes. The morphogenetic mechanisms are robust and yield functional outcomes to a wide variety of inputs. It seems likely that mechanisms responsible for the complex morphogenesis of cells should have special features that make them highly flexible and therefore of advantage to the cells. Like the transcription mechanisms, these features might also make them susceptible to mutation and easy modification in evolution.

A glimpse into the unique biochemical properties of cell morphogenesis came from investigations of microtubule assembly by Tim Mitchison in my laboratory several years ago. This work established that microtubules in the cell turn over very rapidly by means of a novel mechanism, which we called dynamic instability (Kirschner and Mitchison 1986). A typical fibroblast cell has an extensive display of five hundred to one thousand microtubules. Although the population of microtubules as a whole is stable, unless challenged with a drug such as colchicine, the individual microtubules are not; they are continually growing to their full length in about five to ten minutes, and shrink back usually all the way to the cell center in about one to two minutes. Since the microtubule assembly is at steady state, about four times as many microtubules are growing than are shrinking at any time in the cell. Dynamic instability is the stochastic process that causes individual microtubules in a population to shift from a phase of persistent growth to a phase of persistent shrinkage. It is now thought that dynamic instability is an intrinsic property of the microtubule, where assembly of subunits is coupled to its hydrolysis of GTP. Since GTP binding of tubulin is conserved in microtubules in all eukaryotes, dynamic instability must be an important process that exists for a purpose.

Why would the cell continually turn over its microtubules and hydrolyze GTP at a rapid rate, replacing each microtubule one at a time rather than building stable microtubule arrays that can be disassembled at appropriate times, as is the case for nuclear lamins at mitosis? The answer comes in part from the discovery that there are microtubules that are less dynamic than the bulk population, and that these microtubules have specific distributions, particularly in nerve cells as well as in other cell types. Experiments on the morphogenesis of the spindle, and specifically the attachment of microtubules to the kinetochore of chromosomes, suggest that the generation of a stable set of chromosomal microtubules occurs by capturing the ends of microtubules transiently growing out from the centrosome in random directions (Mitchison 1988). Most microtubules depolymerize back to the centrosome, but those captured by the chromosome are stabilized. Thus the function of the rapid turnover of microtubules in the mitotic cell is to generate many different configurations in a short time. The function of the kinetochore is to act like flypaper and stabilize the microtubules that have randomly collided with that structure. The function of dynamic instability then is to generate rapid complexity of structure so that through selective stabilization a functionally appropriate arrangement will emerge.

By comparison to phage or viral morphogenesis, microtubule assembly is both a very simple process involving only one component, tubulin, forming a simple polymer, the microtubule, and a very sloppy process generating and destroying most structures. Yet the enduring value of microtubule assembly is its flexibility. Specific pathways of morphogenesis do not need to be prescribed, but merely a large number of random arrays must be generated with maximum stability of the functionally correct state. Such a mechanisms is highly flexible. Mitosis can start from many initial configurations and overcome obstacles not normally encountered, such as drug-induced depolymerization. As long as the metaphase state is the most stable and the system can move readily from state to state (due to dynamic instability), the spindle will find the most stable (metaphase) state in a reasonable period of time. This requires an input of energy to constantly cause the system to generate many possible outcomes. In phage morphogenesis the pathways must be well described and reliable since the only energy available for an escaping aberrant pathway is thermal fluctuation. Similar mechanisms of microtubule assembly exist in migrating cells, which respond to general extracellular signals and orient toward the signal or in developing nerves (Mitchison and Kirschner 1989). The proper array is most likely to be generated by first stabilizing the ends of microtubules. For cell morphogenesis the final microtubule distribution within each cell will be different, but the general shape will conform to the functional needs of the tissues; in the end there are many functional solutions.

The utility of dynamic instability for adapting and responding to nonstereotyped situations is apparent, and it is not surprising that this property of cells has been so highly conserved. Such mechanisms of obvious advantage to the cells also facilitate evolutionary change. They are robust mechanisms, able to withstand changes in cell properties and still carry out important processes such as mitosis. Changes in cell shape, environmental conditions in the cell, chromosome number, and the presence of other cytoskeletal filaments may all slow down the process but are not likely to cause it to fail. They are also easily modified processes. The conditions that lead to selective stabilization of microtubules are not yet known but, like transcription, presumably could be linked to a variety of extracellular and intracellular signals. The effects of microtubules are far-reaching since microtubules are highly integrative and influence many cell processes. For example, microtubules are intimately involved with vesicle transport, Golgi structure, intermediate filament, and actin distribution. Modifications of the microtubule arrays would therefore influence the overall structure of the cell. Dynamic instability of microtubules has survived a billion years of evolution because of its advantages for cell survival. An associated feature, susceptibility to mutation, may have facilitated evolutionary change. It is a good example of a process that can generate complexity without compromising function.

GTP Hydrolysis and the Generation of Complex Signaling Systems in the Cytoplasm

Microtubules are kept in a dynamic state far from equilibrium through GTP hydrolysis, which allows small perturbations of the system to drive it to the most stable state. It is becoming increasingly clear that GTP hydrolysis is used widely to provide energy in unusual ways, not only to drive unfavorable chemical reactions or provide mechanical work, but to keep systems far from equilibrium so that they can operate with greater precision or greater stability. Such systems are often very adaptable because, unlike phage morphogenesis, they do not settle into a stable state that requires new chemical reactions to reverse. For the same reason, as we shall see, they are also easily modified during the course of evolution.

GTP binding and hydrolyzing proteins were first described in protein synthesis, where their principal role is to increase the fidelity of translation, a process that is not usually subject to regulation. As described in an earlier section on photoreception, heterodimeric GTP binding proteins (G-proteins) are very important in signal transduction, where they couple extracellular signals (e.g., catecholamine, mating factors, light) to internal second messenger systems such as cAMP, cGMP, calcium, or protein kinase C. Several GTP binding proteins are known that couple to different receptors and have different intracellular targets. By expressing different receptors or different GTP binding proteins, the cell can generate complex responses. Recently, intracellular proteins, such as GAP-43 (Strittmatter et al. 1990), have also been shown to couple to G-proteins, demonstrating that the range of interactions of G-proteins is not limited to transmembrane receptors and is likely to be very large.

The cell contains other GTP binding proteins, such as the c-Ras, which plays an important but poorly understood role in regulating cell growth and division. Like the heterotrimeric G-proteins, the activity of the ras protein is governed by the occupancy of the guanine nucleotide binding site by GTP rather than GDP. Auxiliary factors that regulate GTP hydrolysis and exchange have been identified. The number of known GTP binding proteins is growing rapidly (Hall 1990).

An important aspect of cell organization, where GTP binding proteins have a role, is in the proper transport or targeting of vesicles from one part of the cytoplasm to another. In different types of cells this targeting may require movement across an epithelium or in neurons to distant regions of the cell, as well as the conventional problem of moving protein from the ER to the Golgi and on to lysozomes, plasma membrane, and so on. This targeting may have changed in evolution as specialized cell types, such as secretory cells, neurons, platelets, hepatocytes, and endothelium, were developed.

The discovery by genetic and biochemical means that a new class of GTP binding protein is involved in targeting vesicles to various membrane compartments may help explain the simultaneous selectivity and potential flexibility of the processes (Salminen and Novick 1987). Comparing GTP binding protein in signal transduction, protein synthesis and vesicle transport, Bourne has suggested that binding to the proper member triggers GTP hydrolysis and allows fusion (Bourne 1988). Failure to find the right membrane causes the vesicle to diffuse off the site. Specificity is achieved by selection on a population of membranes that collide randomly rather then by a process directing the vesicles along predetermined pathways.

All of these epigenetic mechanisms employ variation and selection on a cell-biological level to generate a complexity of outcomes. Genetic variation and selection among individuals of a population are the two ingredients of the most successful mechanism proposed to account for evolutionary change. Variation also occurs on a genetic level within the immune system of an individual (Borst and Greaves 1987). On an epigenetic level, variation is likely to play an important role in establishing the connectivity of the nervous system, in the morphogenesis of the mitotic spindle, in vesicular trafficking, and possibly in other mechanisms employing GTP hydrolyzing proteins. Such mechanisms are highly adaptive and integrative. For example, a collection of signaling systems that influence the levels of cAMP can only result in a single outcome, a resultant level of cAMP. These signaling systems can reinforce or inhibit each other; each signalling pathway within the cell can add algebraically to give a resultant level of second messenger. The result of these complex inputs is often a specific all-or-none decision, for example, whether to grow and divide, whether to differentiate, or whether to signal another cell. Such mechanisms are also very susceptible to mutation. It is easy to see how new linkages would drastically change cell behavior as a result of mutation. The expression of new selective components such as adhesive molecules in the nervous system, receptors or modulators in the signaling system, or stabilizing or destabilizing proteins in the cytoskeletal systems could result in wholly new responses for cells. By small mutational changes organisms could become stereotyped or flexible in cellular functions.

Evolution of Developmental Mechanisms

We have seen that evolution occurs on the level of gene products and on the level of ensembles of interacting proteins. Although the appearance of new genes can be crucial for major new forms of life—as, for example, scales and feathers in birds and reptiles—it is usually the case that the linking features in evolution are the pathways of linking these functions together. Thus more complex systems such as the visual system or the skin involve many gene products—some old, some modestly remodeled, and some newly brought together for a specific function.

On the larger scale of the development of organisms there has also been an evolution of collections of cellular mechanisms characterized by flexibility, sensitivity to mutational change, and robustness in giving many functional responses. These mechanisms carry within them the capacity to generate easily many different morphologies and activities, often without gross disturbance to the organism. These collections of cellular mechanisms can be called developmental mechanisms, and these developmental mechanisms, like their constituent cellular pathways, have also evolved, selected for their capacity to sustain without loss of function.

Three examples of important developmental mechanisms are head neural crest formation, germ band formation in insects, and gastrulation-neurulation in vertebrates. Although each of these is a morphological term or stage in embryogenesis, it is not really the anatomical descriptions that are important, but the fact that they represent a conserved collection of cellular mechanisms with rules for cell interaction and specific patterns of cell behavior.

The neural crest first appears in vertebrates; it is absent in cephalochordates like amphioxus, which do possess a notochord, somites, and neural tube. The head of vertebrates has undergone considerable enlargement and evolutionary change. Much of the tissue in the head is of ectodermal origin. The tissues include brain, which arises from the neural tube, and the bone, cartilage, muscles, and skin of the head, much of which arise from the neural crest. Many of the tissues that are typically of mesodermal origin in the trunk are generated in the head by the neural crest, suggesting that a radically new developmental program underlies vertebrate head development (Gans and Northcutt 1983). The neural crest cells divide rapidly and migrate as single cells. They then coalesce to form peripheral nerves, muscles, and bones. The capacity of the neural crest to migrate as single cells frees it from the morphological constraints of forming muscle in the trunk through the stereotyped morphogenetic changes of the somite tissue. In looking at the extensive radiation of the vertebrate head, Gans (Gans and Northcutt 1983) has credited the multipotentiality of the neural crest, its proliferative capacity, and its plasticity for the rapid evolutionary change of the head. Neural crest formation, as a process, must be one of the great innovations of vertebrate development. It represents not a single gene or a single pathway but a collection of mechanisms capable of generating great complexity, as well as capable of being easily modified. Later in development, wholly new structures emerged from head neural crest—for example, the reorganization of the branchial arches into the jaw (Langille and Hall 1989). The neural crest tissue, having evolved in vertebrates, underwent rapid evolutionary change, replacing the traditional mesodermal tissues of many cell types.

Some developmental mechanisms have changed very little in evolution. It is curious that this is so, especially when earlier stages and later stages have undergone many more evolutionary changes. The germ band of insects and the neurula stage of vertebrates are two examples, where the early embryonic events that led to this conserved morphology have undergone a great deal of evolutionary change but the structure ultimately formed is very similar. For example, the early cleavages and gastrulation movements of fish, amphibians, and mammals show widespread differences. In the case of fish, cells from the ectoderm migrate and aggregate to form a mass of cells that generate the neurula embryo. During this process the cells may lose any lineage relationship (Kimmel 1989). Neurulation in fish occurs by cavitation and without a folding of the neural epithelium, while in amphibians, birds, and mammals the epithelium folds. In most amphibians gastrulation proceeds by involution of sheets of cells that move as a coherent unit. In birds and in very yolky amphibian eggs that show meroblastic cleavage, the invagination of sheets of cells is mechanically impossible and cells migrate singly through a primitive streak. In mammals the early cleavage division serves a novel purpose, the formation of extraembryonic tissues. Despite these highly disparate cellular mechanisms in the early embryo, all of these processes lead to a highly conserved structure, the postneurula stage, which Ballard has called the pharyngula (Ballard 1981). The pharyngula of all vertebrates is very similar. It is made up of a dosal hollow nerve chord, notochord, segmented somites, and pharyngeal slits. To this basic structure of all chordates, vertebrates have added distinct structures of neural crest origin (Gans 1989). The basic body plan has been retained and is the basis for an incredible amount of radiation.

The story is very similar for the germ band of insects. This structure incorporates the body plan and segmental organization of all insects. Molecular markers show that both are similar: for example, they contain fourteen domains of the engrailed transcription factor in a parasegmental pattern (Akam 1987). The formation of the germ band in short germ band insects, such as the grasshopper, is very different from its formation in long germ band insects such as the fruit fly. Extensive recent studies have shown that the fly generates the anterior-posterior pattern in the larva by sequentially partitioning the entire embryo within a common cytoplasm. Starting with prelocalized information in the anterior and posterior regions, boundaries are set up incrementally within a common cytoplasm. Most of the signaling molecules are transcription factors that can diffuse among the nuclei in the syncytial embryo. Short germ band insects develop quite differently. The middle and posterior segments are generated not by partitioning of a common cytoplasm but by cell proliferation. There is no opportunity to diffuse intracellular transcription factors over the entire anterior-posterior axis, since the posterior segments are not partitioned but generated by new cell division. Yet the same germ band structure forms with its parasegmental stripes of the engrailed protein (Patel et al. 1989).

The pattern of development in insects and vertebrates does not correspond to Von Baer's plausible idea that the earliest stages of development should be most conserved and that evolution should proceed by adding refinements later in the ontogenic process. Instead we find major divergence in the earliest stages, conservation in the middle, and divergence at later stages of development. The early divergence suggests that early developmental mechanisms are not ummutable and the conservation of the pharyngula and germ band stages is highly significant. It seems most likely that structures like the germ band and pharyngula have served and continue to serve an important function, and that they represent the anatomical manifestation of a set of cellular mechanisms that can give rise to many developmental solutions (for example, all of the vertebrate and insect forms). As such, these grand developmental mechanisms—like the simpler cellular mechanisms of transcription, cytoskeleton, and cell signaling—are engines of variability. Like the cellular mechanisms, these developmental mechanisms are robust and easily mutable to give new functional outcomes. However, unlike the cellular mechanisms, it will be difficult to identify specific genes responsible, since these developmental mechanisms are complex collections of many cellular processes involving, for example, extracellular signaling molecules, cell surface molecules, internal signaling molecules, and transcriptional cascades. From this collection of mechanisms emerge morphogenetic movements, sorting out, and new tissue specifications.

I presume that the developmental mechanisms themselves, just like the cellular mechanisms, must have evolved, which means that they must have been, and perhaps still are, genetically variable among individuals. Yet paradoxically their extreme conservation suggests that, once established at an early date, they were difficult to change. Nevertheless, when organisms perfected these plastic and adaptive developmental mechanisms they were also facilitating further evolutionary change. Successful new developmental mechanisms such as the neural crest or vertebrate limb bud not only provided a solution to the immediate adaptive problem but generated mechanisms that were well suited for later rapid modifications. The opposite must also have been true. The failure of organisms to develop such mechanisms would have limited their radiation by imposing what is referred to as developmental constraints.

Although the hardware of new gene products is important, as is the software of new cellular mechanisms, the truly great moments in evolution came when operating systems, or collections of software, were created. These developmental operating systems have evolved the capacity to generate new software and to make use of new gene products. Aside from the neural crest, gastrulation, and the germ band, there are many other similar developmental mechanisms, such as the trochophore larva and the vertebrate limb bud. They have evolved as dynamic structural assemblages with a tremendous capacity for evolution. They have evolved not as anatomical units but as a set of biochemical circuits involving diffusible factors, receptors and linkages to intracellular signaling systems, and the cytoskeleton. These developmental operating systems can be easily altered by mutation and have a propensity to yield functional solutions. It is the study of these developmental operating systems that, in my opinion, offers us the best opportunity for understanding the mechanisms of evolution. With an understanding of these systems will come an appreciation of why eukaryotic cells are the way they are.

John Bonner's work with slime molds (e.g., Bonner 1977) has demonstrated that the study of development can tell us much about cell biology, and that the study of development can also raise new questions about evolution. I hope cell and molecular biologists will soon be in a position to return the favor, and that molecular and cell biology will be able to inform us about the mechanics of development as well as the mechanics of evolution itself.

Summary

Relatively little effort has been made to trace evolutionary advance on a cellular level. Within eukaryotic cells recent evidence has suggested an unexpectedly high level of conservation among cellular components from phylogenetically distant organisms. While this has reinforced our confidence in the evolutionary relationship among organisms, it has not clarified the important changes on a cellular level that underlie the diversification of organisms, nor has it indicated the important cellular mechanisms that must underlie evolutionary change. In this essay I look at the nature of cellular changes in evolution. Evolution is best understood on the level of evolution of cellular processes. Existing components may be merely reshuffled and reorganized to give new functions. Novel gene products do arise and make important contributions, but their contributions depend on the organization of new processes, which may come considerably later than the new gene products. Integrative cellular mechanisms, such as cell-signaling pathways and transcription, themselves have evolved with properties that in turn facilitate evolutionary change. These mechanisms have the capacity of generating many functionally different outcomes with little mutational change. Collections of cellular mechanisms, which I call developmental mechanisms, such as gastrulation, have also evolved, with the same end: generating with few mutational changes, divergent yet often functional outcomes. With this perspective we may be able to identify the important cellular changes that underlie evolutionary change and, although we cannot predict the course of evolution, we can at least understand evolution as a biological process.

Acknowledgments

I thank John Gerhart for his active participation in the formulation of the ideas in this chapter and for his comments on the manuscript itself. I am grateful for valuable conversations with J. Brown, H. Gainer, K. Yamamoto, and I. Herskowitz. And finally, I thank James Sabry and Tim Stearns for their comments on the manuscript.

7

Will Molecular Biology Solve Evolution?

MARTIN KREITMAN

More effectively than anyone of his era, John Bonner's research spanned the two great experimental traditions in biology: cellular and developmental biology on the one hand, and organismal and evolutionary biology on the other. Although he recognized certain essential differences between them—the kinds of questions evolutionists ask are different than those of cell biologists—he nevertheless spent much of his career trying to minimize the distinctions.

In *On Development* (1977), for example, he espoused a view of life bordering on the metaphysical: all biological processes, whether they be molecular, genetical, cellular, developmental, organismal, or even populational, are manifestations of one simple governing principle, the life cycle. DNA replicates through cycles of semiconservative replication, enzymes cycle through endless rounds of synthesis and degradation, organelles obey the cell cycle, cells cycle through development, organisms replicate just like DNA but through the generation cycle, and populations cycle from one generation to the next through a sieve we call natural selection.

This self-contained conception, by entailing all levels of biological organization, provides a unified framework for viewing all of life's processes, from molecular reactions to behavioral evolution. Under this view, development and evolution are two interconnected parts of a grand life cycle that carries individuals (and populations) from one generation to the next. Development produces adults from eggs; natural selection cycles certain chosen individuals from one generation to the next.

This book is concerned with the future of evolutionary biology. Here, I would like to consider a practical side of the issue connecting development and evolution, namely, how to (or not to) exploit molecular approaches for evolutionary purposes. More precisely, I would like to ask: Is there a common intellectual meeting ground for evolutionary biology and modern molecular biology? I do not intend to address here what I view to be essentially nonscience issues—hinted at in John Bonner's struggle to formulate a functional worldview for biology—whether, for example, molecular, cell, and developmental processes can be fully understood without the light of evolutionary theory; or conversely, whether our understanding of evolutionary mechanisms requires knowing the rules of development. For the purpose of this essay I will assume evolutionary thoery to be adequate for the purpose of formulating well-posed questions about evolution at the molecular or cellular level. And no evolutionist can call into question the importance of understanding development. Rather, my purpose here is to address a far more utilitarian question: Is the molecular approach as likely to succeed in evolutionary research as it has in medicine, genetics, and development? What are the limits of the molecular approach in solving evolutionary problems?

Whether the molecular approach has a future in evolutionary biology is relevant at another level: the future of evolutionary biology research in the academic life sciences. Many universities, in order to accommodate two very different research modes in biology, have separated suborganismal and organismal research among two departments, Molecular Biology and Ecology and Evolution. Reconciling (or maybe recognizing) the expanding gulf between the two research traditions, intellectual foci, and funding trajectories, I explore in this chapter whether and where bridges are likely to form.

I take on this task for a number of reasons, but primarily because my own work uses molecules to study evolutionary processes. Molecular evolution is thought by many to be the new span that links the two worlds, and I want to articulate here why I disagree with this view. It is more than just a coincidence that neither conceptual nor organizational structures have been capable of preserving the unity of biology: evolutionary biology and molecular biology have become fundamentally different sciences, increasingly pursued in separate departments within universities. If there is any meeting ground, I believe it consists of the many one-dimensional points of departure, such as "DNA," rather than multidimensional fields of inquiry, such as "development." Indeed, I do not think the two research traditions have much at all in common, and to the extent one tradition informs the other, the flow of ideas is neither symmetrical nor synergistic.

Recent discoveries in the molecular realm have been revolutionizing our ideas about the nature of gene expression and the mechanisms of development, but the same discoveries will not necessitate major revisions of our evolutionary theories. Knowing the molecular and genetic bases of development, for example, may help us to distinguish between competing explanations for patterns of morphological evolution; but nothing we learn about evolution from development will demand explanations falling outside of the modern evolutionary paradigm. New discoveries at the molecular, cellular and developmental levels will, undoubtedly, enrich our understanding of evolution and the mechanisms of change (be it morphological, physiological, or developmental), but these discoveries will lead to no new paradigm shifts: evolution by natural selection and genetic drift acting on individual differences within populations will survive as the principle driving forces in evolution.

Evolutionary biology, in my opinion, has placed naively unrealistic expectations on the molecular approach. There are limits to how precisely we can apply the tools of molecular biology to evolutionary problems, however precise the measurement tools; and there are limits to how effectively we can apply the experimental *sine qua non* of molecular biology—reductionism and strong inference. In this chapter I will elucidate some of the reasons that have forced me to adopt this rather severe view.

My First Encounter

As a new graduate student, my very first day of exploration in the evolutionary biology stack of the library led me to a rather new and silvery book called *On Development*, by an author whom I somehow knew to be famous. Imagine, a book on cell and development with its first half devoted to evolution and natural selection! I took it to a table, sat down, and started reading. I can honestly say that it remains the only science book I have ever read in one sitting. And I distinctly remember what an enjoyable experience it was. The freedoms of graduate school were going to be greatly rewarding.

But I also distinctly remember struggling with a rather disturbing feeling, having just completed the book, that as enjoyable as it was, and however informative I found certain parts to be (the biology of slime molds in particular), I did not comprehend the proposed union of evolutionary biology and molecular development. The thought disturbed me because this idea of a deep connection was a central premise of the book. I recall thinking possibly the book was too rich to digest in a single sitting, and so I left with the hope of taking another look at it sometime in the future.

That second look came in preparation for this essay. The first feature of the book that struck me was its symmetry: its 258 pages divided into two sections: (1) The Molecular Basis of Evolution, and (2) The Molecular Basis of Development, with exactly 129 pages devoted to each topic! My initial impression was correct: there was no union. Evolution and development, equal but separate. It was as if John Bonner was cementing in place two bulkheads to be bridged at a later date by someone else.

Indeed, I think this is true of much of his work; he identifies interesting features of the biology of an organism, separately considers phylogenetic trends relating to the particular biology, insists there must be an evolutionary connection, and takes a stab at a solution. He is iconoclastic only in insisting on a Darwinian solution. I will return later to the issue of whether or not we can ever know if the evolution of development and morphology is indeed Darwinian.

The Bridge Connecting Molecular and Evolutionary Biology Contains One-Way Traffic

The flow of information between molecular biology and evolutionary biology is not symmetrical. Scientific reductionism is the guiding principle for all modern experimental biology, molecular and evolutionary. But as Peter Grant has pointed out to me, in scientific reductionism the flow of information is primarily unidirectional, flowing successively upward from the most microscopic levels of biological organization to the more macroscopic organismal levels. For example, biochemistry might be expected to inform physiology but not the other way around.

The same is true at the higher levels of biological organization. For example, population genetic theory rests on certain basic assumptions about gene segregation and assortment. The principles of gene segregation and assortment, in turn, derive from knowledge of chromosome behavior in meiosis. But it would be absurd to claim that our understanding of meiosis is enriched by results from population genetics. Evolutionary biology is informed by molecular, cell, and developmental biology, but not the other way around.

Molecular biology appears to have no need for evolutionary biology and the absence of need explains the information-flow asymmetry. Lip service is occasionally paid to the value of evolutionary biology, but rarely are words translated into deeds. I once heard David Baltimore, the Nobel laureate and then head of the Whitehead Institute at MIT, having discovered the tremendous conservation of genes and genomes, proclaim that to be a good molecular biologist one would also have to be a good evolutionist. Despite that MIT remains MIT, the very model of a molecular biology leviathan, with its institutionalized adherence to scientific reductionism.

Another common manifestation of the one-way flow of information reveals itself in my daily interactions with molecular biology colleagues. A colleague will say something like, "My lab just discovered a new gene in yeast (or organism X) that turns out to have homology to such-and-such a gene in the fly (or organism non-X). You have to explain to me just what this means." Of course, I have nothing useful to say to them. Even if I did have a novel evolutionary insight, it would unlikely be one that would bring them any closer to understanding of their gene. On the other hand, I am selfishly happy to hear about their new discoveries because I feel it might help me understand molecular evolution. So, to keep the information coming, I (try to) placate them with another application of some well-worn evolutionary dogma. Fortunately for me, many of my molecular colleagues have short memories for evolutionary dogma.

Although molecular biology does not need evolutionary biology for their work, without evolutionary explanation their work lacks intellectual breadth and remains disconnected from the historical process that explains the generality of their findings. Historical contingency is demonstrated again and again in molecular biology (consider the evolution of duplicate genes and novel gene functions); all of biology gains from this knowledge. The challenge of evolutionary biology is to solidify our knowledge of phylogeny and process—the complementary structures onto which all molecular findings must be placed in order to be fully understood, and to convey the intellectual purpose of this endeavor.

Revolution and Evolution in the Two Biologies

I view the two biologies as being at two distinctly different phases of their intellectual growth. Molecular biology is almost completely a science of the second half of the twentieth century. Although one might argue that certain fields of inquiry, such as embryonic development or inheritance, are as old as biology, I would say that more has been learned about gene regulation and development in the past ten years than in practically all the preceding years. Unlike evolutionary biology, molecular biology does not need its past: tomorrow's discoveries depend on little more than today's results. We evolutionists simply have to recognize the remarkable scientific revolution being brought about by molecular biology.

Evolutionary biology, in contrast, rests on a conceptual foundation laid by Darwin over a century ago. Most of the theory's edifice, including its genetical wiring, was completed in the first half of the twentieth century by a small number of visionary scientists, led by R. A. Fisher, S. Wright, and J.B.S. Haldane. In my view, only two substantive wings have been added to the theory in the second half of the twentieth century: W. D. Hamilton's (1964) theory of inclusive fitness for understanding social evolution, and M. Kimura's (1983) development of the neutral theory for understanding molecular evolution.

Therefore, today's evolutionist is concerned mostly with detail work: Darwin's theory is conceptually well formulated and empirical support for the general theory is unassailable. My question is: Is it only a matter of time before evolutionary biology turns over the maintenance job to philosophers and historians? (Consider this: William Provine, the noted historian of science, already teaches the general Evolutionary Biology course at Cornell.)

I have tested out my idea that evolutionary biology is near its endgame (which I believe only half-heartedly) on a number of colleagues and have received only two kinds of responses. The first rejects the notion that the modern evolutionary synthesis entails the correct mechanisms for explaining the history of life. The other, while recognizing the breadth and depth of our conceptual understanding, points to the lack of well-worked-out empirical examples.

Stephen Jay Gould and Niles Eldredge's theory of punctuated equilibrium is a good example of a recent challenge to modern evolutionary theory (Eldredge and Gould 1972). The theory proposes new levels of selection to explain patterns of taxon evolution. Populations are seen as only one of many hierarchical arenas in which evolution occurs, each with its own rules and regulations. But this theory has been heavily—and many people feel adequately—criticized by defenders of orthodoxy. It is unclear whether certain kinds of species selection are simply reifications of populational Darwinian selection to a higher level.

Another challenge to conventional evolutionary theory comes from Leo Buss (see Chapter 5; see also Buss 1987). In trying to understand the evolution of complex life cycles in marine invertebrates, he challenges the notion that the individual is the unit-of-selection. He insists

Darwinian evolution cannot explain the occurrence of complex life cycles. Standard theory, according to him, accounts for only a very small component of evolutionary change. He considers evolutionary theory to be in its infancy. I find this idea both easy to reject and difficult to refute. Evolutionary theory is undoubtedly incomplete, but, taking molecular evolution as but one example, many (but not all) observations can be understood in the framework of modern evolutionary theory. The burden of proof, it seems to me, is on the challenger.

An interesting exception to the idea that nothing new has occurred in evolutionary theory comes from the observation that the genome contains a great deal of repetitive DNA. Copies of a repetitive DNA sequence at different sites, or loci, in a genome are often more similar to one another than are true homologs residing in different species. This should not be so if the sites within a genome were evolving independently and their origins predate the species split. To explain the evolutionary nonindependence, or *concerted* evolution, G. Dover has proposed a new (and distinctly non-Darwinian) evolutionary mechanism for the horizontal spread of new variation, called *molecular drive* (Dover and Tautz 1986). Whether or not this mechanism can lead to the fixation of nonadaptive mutations, as claimed by Dover, remains the subject of some debate. Although the relative importance of molecular drive in evolution is debatable, the discovery of a new mechanism for evolution is not.

The second kind of challenge to my near-senescence conjecture is that as long as there are biologists around, even if only of the molecular persuasion, their new discoveries will require evolutionary explanation. So, for example, the discovery of a separate genome for mitochondria demands an explanation for how it arose. In response, we now have the endosymbiont theory for the cellular acquisition of organelles. One can take this kind of argument only so far. Certainly it is not the obligation of evolutionary biology to instantiate with evolutionary explanation every single observation in biology. Indeed, evolutionary biology already suffers from a fixation on explaining life's many singularities at the expense of explaining the commonalities.

Molecular biology has adopted a rather different tack, and it is one we evolutionists should better appreciate. The core discoveries in molecular biology have all been made on only a couple of viruses and a half dozen animals, *Escherichia coli, Saccharomyces cerevisiae, Caenorhabditis elegans, Drosophila melanogaster, Mus domesticus*, and *Homo sapiens*. Although it need not have turned out this way, there is a remarkable evolutionary conservation of biochemical and developmental pathways. Consider, for example, DNA binding proteins controlling gene expression. Who would have believed that the three-dimensional structure of the helix-turn-helix DNA binding motif (see frontispiece) of the following proteins is virtually undistinguishable: the homeodomain of *Drosophila*, phage 434 repressor, lambda repressor, phage 434 Cro protein, the Trp and Lac repressors of *E. coli* (Gehring et al. 1990)?

Although evolutionary biologists are quick to point out molecular biology's failure to recognize its own limited view of life, so evolutionary biology has not sufficiently developed and exploited a few model organisms.

My own view about the status of evolutionary biology is that what remains to be solved are not at all new problems, but instead mostly old ones that keep coming back to haunt us. This is certainly true for my own research on the forces maintaining genetic variation in natural populations. I am little more than a second-generation descendant of T. Dobzhansky (his F_2, as it were), who defined the problem some sixty years ago.

Many of the simplest and most common patterns of morphological evolution still elude sastisfatory evolutionary explanation. In a recent conversation, Richard Lewontin pointed out to me that many morphological characters are essentially invariant within species—the scutellar bristle number and position in *Drosophila*, for example—but are manifestly different between species. The "whole problem of evolution," according to him, is to explain this seeming contradiction. Why, he wants to know, do characters like that exist? Such a view should warm John Bonner's heart. Indeed, there are enough classical problems still around to keep us evolutionists in business for some time to come.

Limitations of the Molecular Approach to Evolution

For the reasons outlined above, I do not believe evolutionary biology and molecular biology will emerge anytime soon as a unified science. (One encouraging exception to this claim is the emergence of a field of study involving the evolutionary analysis of gene and genome structure.) But at the same time, molecular approaches have become an integral part and parcel of the evolutionary biologist's tool kit.

I will now focus on issues having to do with limits to what we can expect to learn about evolution from molecules. What can evolutionists expect from the molecular approach in the last decade of this remarkable century of discovery? For example, will molecular biology do for our understanding of speciation what it has already done for gene regulation? I cannot provide a timetable for discovery. Molecular biologists cannot even predict the progress of their own science—for example, the time to a cure for cancer, or a date for a completed human genome sequence. Similarly, no one can predict when our understanding of developmental genetics will be sufficient to allow us to know how development constrains evolution. But I can devote the remainder of this chapter to identifying some of the limitations of the molecular approach to evolutionary problems. Too much, in my opinion, is being expected of the molecular approach and too little consideration is being given to the epistemological constraints in evolutionary biology.

The limitations of the molecular approach fall, conveniently, into four categories: technical, theoretical, epistemological, and economic/political. In short, I believe that technical innovations will be introduced at least as fast as evolutionists can apply them; theoretical considerations impose moderate limits to making evolutionary inferences from molecular data; epistemological constraints are more severe than generally recognized; and financial/political factors are near fatal.

Technical Limitations

Although molecular biology continues to get easier and more accessible, it remains time-intensive, expensive, and technically frustrating for newcomers. But now, at least in work involving the genetic analysis of individual or populational differences, we have more information than ever before. First, there is a remarkably high level of polymorphism at the DNA level. Our best guess for *D. melanogaster,* for example, is that one nucleotide site in about every two hundred is segregating in an "average" diploid individual (Aquadro 1989; Kreitman 1990); for other species of *Drosophila* it may be as high as one in fifty (Riley et al. 1989). As to polymorphism levels in non-*Drosophila* species, to a first approximation we expect the amount of nucleotide polymorphism to be directly proportional to the evolutionary effective population size. The *Drosophila* polymorphism levels are consistent with minimum population sizes in the range of 10^6–10^7. As for other species, mammals are undoubtedly less polymorphic than *Drosophila,* but the opposite may be true for many marine invertebrates.

Species with a long generation time, like humans, may turn out to have surprisingly high levels of polymorphism, belying their small evolutionary effective population sizes. One estimate of heterozygosity in humans, based on restriction fragment length polymorphism, is one in five hundred base pairs, a factor of only two or three lower than for *D. melanogaster* (Antonarakis et al. 1988). If this estimate turns out to be accurate, the explanation may involve the huge generation-time difference between fly and human. If mutations occur at a constant rate per year, rather than per generation (this is a very important and unresolved question), then the per generation mutation rate in humans would exceed the rate in the fly by at least a factor of one hundred. This would bring the heterozygosity estimate for humans, with a 10^4 evolutionary population size, into line with the heterozygosity value for the 10^6 population size of *D. melanogaster.*

However, a more recent report, based on DNA sequences of coding regions of approximately sixty different genes, suggests a much lower estimate of human polymorphism, approximately one base in ten thousand (Li and Sadler 1991). Now, if this value is correct and population size is correspondingly smaller than in *Drosophila,* then, as Li and Sadler point out, a new conundrum arises: why are allozyme polymorphism levels in *Drosophila* and humans approximately the same? Li suggests a model in which most amino acid replacement changes causing polymorphism are slightly deleterious. In small human populations, many more of these slightly deleterious mutations will behave as effectively neutral alleles, whereas in large insect populations they are definitely selected against. This effect cancels out the higher intrinsic polymorphism levels expected in larger population size species if polymorphisms are completely neutral. Of course, there is an alternative to the slightly deleterious model: heterozygote advantage. It is doubtful, though, whether balancing selection is ubiquitous enough to explain such a general pattern as overall protein polymorphism levels. Unfortunately, we still do not know the answer to this most fundamental evolutionary problem!

Nucleotide substitution is not the only kind of polymorphism, and some other kinds have considerably higher mutation rates and hence polymorphism levels. Here there are two kinds of useful length variations: the immensely variable nuclear tandem repeat sequences, VNTRs, such as the ones first identified by Jeffries, which can be so variable as to have hundreds of alleles segregating at a single "locus" (Jeffreys et al. 1985; Gyllensten et al. 1989); and the simple dinucleotide tandem repeats, CA/GT, of which there are estimated to be approximately 100,000 in humans and which can segregate for tens of alleles per "locus" (Weber and May 1989). These polymorphisms are already being exploited in evolutionary biology in the study of population structure, for paternity or maternity analysis, and as markers for gene mapping.

If the nuclear genome does not provide enough evolutionary ammunition, there is always the maternally inherited mitochondrial genome in animals and the chloroplast genome in plants. Curiously, whereas the mitochondrial genome tends to be quite polymorphic and hence useful for populational studies (Cann et al. 1987), the chloroplast genome is exceedingly slow-evolving, which makes it useless for population studies but exquisite for plant systematics (Palmer 1982).

As to methods for finding variation, molecular biology is now practically vectorless, thanks to PCR, the polymerase chain reaction (Sakai et al. 1986). DNA sequencing is also continually improving, albeit slowly. But with the human genome projects now committed, we can expect to see an order-of-magnitude improvement within the next ten years. In summary, while molecular biology remains technically demanding, there are plenty of readily available tools for use by evolutionary biologists, and there is no shortage of variation to exploit.

Theoretical Limitations

For many kinds of evolutionary problems there are definite and often calculable limits to the applicability of a molecular approach. Although many of the limits are not severe, every one of them should be investigated in advance. Here are a few examples, drawn primarily from my own area of expertise, population genetics.

At any one genetic locus, say the protein coding region of a particular gene, many different nucleotide sites are likely to be variable within a population. With a heterozygosity of 0.005, as in many regions of *D. melanogaster,* a diploid individual is expected to have ten segregating sites in every two thousand base pairs. Given enough recombination between polymorphic sites, the ten variable sites can produce as many as 2^{10} (1,024) different allelic combinations or haplotypes. But because most molecular techniques for identifying polymorphism, such as Southern blot analysis, do not allow us to distinguish which chromosome is segregating for which polymorphism in diploid individuals, it is impossible to deduce haplotypes from the genotype of an individual. All this potential information is lost.

In a similar vein, even if one knew the haplotypes of a set of alleles, as would be the case if the complete DNA sequence of each gene were known, it would still be practically impossible, because of recombination, to determine the genealogy (or phylogeny) of the alleles. The problem of knowing which segments of which sequences are recombinationally rather then mutationally related is, as far as I can tell, nearly hopeless. Therefore, although all alleles are genealogically related, all having descended from common ancestral sequences, this evolutionary information may not be recoverable!

DNA does not allow unlimited precision of estimation for many ecological and evolutionary parameters. In population genetics, the estimation of heterozygosity, a fundamental evolutionary parameter, has a large uncertainty (the variance is expected to exceed the mean) and the variance decreases linearly with the log of sample size (Ewens 1979). Therefore, to double the precision of an estimate of heterozygosity for a region of DNA, we must increase the sample size by a factor of ten. Improvements in technology will undoubtedly allow us to gather data from ever larger samples, but they will never overcome this theoretical obstacle.

The same is true for the analysis of relatedness based on hypervariable VNTRs. Whereas the method of matching genotypes is useful for determining very close genetical relatedness, such as maternity or paternity, the probability of exclusion is expected to decrease exponentially with decreasing genetic relatedness (Lynch 1988). So for all practical purposes, exclusion analysis is good only for determining parent-offspring and sib-sib identities. Even identification of cousins is unlikely.

Probably the broadest application of the molecular approach in evolutionary biology is for phylogeny reconstruction. But there are certain limitations here as well. They are partly imposed by DNA itself—there are only four possible character states, T, C, G, and A, at any nucleotide site, and backmutations are expected to be increasingly common as the time from a common ancestor increases. Other limitations are imposed by the problems associated with finding the true phylogenetic tree. Those who think that DNA, with its many possible characters (nucleotide sites), obviates the latter problem should read Felsenstein's two excellent reviews of tree-building methods (Felsenstein 1982, 1988).

One of the most interesting applications of molecular phylogenies has been to the question of the major and most ancient branch points of life.

For this work, only the slowest evolving regions of DNA can be useful. (Noncoding regions, for example, may be mutated to an unalignable state within fifty million years or so. And they are useless for phylogeny long before that.) Most of the ancient phylogeny reconstructions are based on changes in the highly conserved 16S/18S ribosomal DNA subunits. But are the rare mutational changes at very slow-evolving genes good characters for phylogenetic reconstruction? There is no reason, a priori, why they should be, given the potential for every nucleotide site in every genome to be mutated as frequently as once every one hundred million years.

Mutations can be thought of as coming in only three varieties: those that are selectively advantageous, those that are neutral, and those that are deleterious at any one time. In slow-evolving genes, by definition, the vast majority of mutations must fall into the deleterious category; it is the only way these genes can be evolving so slowly. This being the case, why do these genes evolve at all?

If the rare mutational changes of slow-evolving genes are selectively neutral, then for species that have been separated by hundreds of millions of years there will have been time for a neutral site to have mutated back and forth many times in each lineage. Such changes would not be particularly reliable as indicators of phylogeny.

If, on the other hand, mutational substitutions at slow-evolving genes are adaptively favored, then no matter how small the possible number of adaptive mutations, there will also have been time for the same parallel adaptive change (homoplasy) in more than one lineage. Again, they would not be particularly reliable as indicators of phylogeny.

The third possibility is that all mutations are deleterious but that very rarely one nevertheless fixes in a species by genetic drift. But again, I would argue, adaptive evolution will have had sufficient opportunity to reverse the course of this deleterious evolution. There only needs to be sufficient time for the appropriate backmutation to occur. By definition, such a mutation will now be selectively favored.

The only mechanism for slow evolution I can think of that would be sympathetic to ancient phylogeny reconstruction is if one mutational substitution, whether it be selectively favored or deleterious, changes the fitness consequence of mutations at other sites. This change might provide a mechanism for producing distinct molecular evolutionary trajectories in different phylogenetic lineages. But it is doubtful whether this is the only mechanism for slow molecular evolution. Slow-evolving genes may preserve some phylogenetic information, but I suspect the information is contained among many false leads.

Reconstructing the tree of life is one of the most exciting evolutionary applications of molecular biology, and although tremendous progress has been made (see Lake 1991 for a recent review) this is a subject that deserves serious theoretical attention. But at least if, for whatever reasons, highly conserved molecules do retain ancient phylogenetic information, the same phylogenies should obtain using different molecules.

Phylogenetic consistency, however, is not expected at the opposite end of the phylogeny reconstruction spectrum: resolving the relationship of very closely related species. Our own species contributes to one such example, the (unresolved) trichotemy of human, chimp, and gorilla. With very closely related species, the time separating the species may be insufficient for the species to have accumulated fixed mutational differences between lineages. Most differences will then represent the sorting out of variation already present in the common ancestor of the species. In such a case the genealogy of the alleles may be a poor indicator of the phylogeny of the species. One dramatic example is documented in *Drosophila simulans*, a geographically broadly distributed species, and two very closely related island endemics, *D. sechellia* (Seychelles Islands) and *D. mauritiana* (Mauritius Islands). In these species, almost all the differences between species result from the segregation of ancestral polymorphism (Coyne and Kreitman 1986). In such a case, many loci and many individuals per species will have to be compared in order to obtain a consensus phylogeny.

Epistemological Limitations

The two biologies ask very different kinds of questions about organisms, and they require distinctly different methodological and experimental approaches. Very simply put, cell and molecular biologists are interested in knowing how an organism and its component parts work, whereas evolutionary biologists want to know how the organism got to be the way it is. Of course there is overlap: a whole branch of evolutionary biology deals with functional anatomy. But in a very general way, molecular biologists are primarily deconstructionists, evolutionists are reconstructionists.

Whereas molecular biology relies exclusively on scientific reductionism and strong inference for its discoveries, evolutionary biology has always required more indirect approaches. One reason for the difference is purely practical: it is substantially easier to experimentally manipulate populations of molecules or organelles in a test tube than it is to manipulate populations of organisms, or worse yet whole communities of organisms, in nature. The analysis of variance is often the best we can do in evolutionary biology, not because it is the approach of choice, but because there *is* no other choice. Although the two fields borrow from each other—molecular biologists identify functionally important regions of genes by the comparative method (the *sine qua non* of evolutionary biology), and certainly evolutionists strive to perform critical experiments—evolutionary biology will never be as exact a science as molecular biology.

It might be argued that evolutionary biology has more tools available to it than simply the analysis of variance. After all, are not the best studies in evolution ones that combine natural observation and measurement with direct experimental confirmation? For example, in a replicated experiment Reznick, Bryga, and Endler (1990) reciprocally transplanted guppies in nature into high and low predation environments. Over a period of years they were able to demonstrate experimentally a shift in the reproductive schedule of the transplanted populations through the actions of natural selection. The shortened reproductive schedule in the high predation environment and the lengthened schedule in the low predation environment mimicked what they observed in undisturbed populations.

But there are also experimental situations in molecular population genetics in which no amount of replication or experimentation will always resolve the workings of natural selection. Consider the most sensitive system available for detecting selection in an experimental setting, the competitive growth of the bacteria *Escherichia coli* in chemostats. In a classic experiment, Dykhuizen (1978) was able to demonstrate by direct-competition experiments the selective advantage of tryptophan auxotrophs over prototrophs in chemostats when the amino acid tryptophan was supplied in excess. His results are shown in figure 7.1. But he was also able to show that the selective advantage of the auxotroph was not related to the energy savings of not producing the unneeded Trp enzyme. The biochemical or energetic mechanism for the selective advantage of the Trp auxotroph remains unexplained.

A much more severe problem with chemostat experiments was revealed in an experiment to test the selective neutrality of four naturally

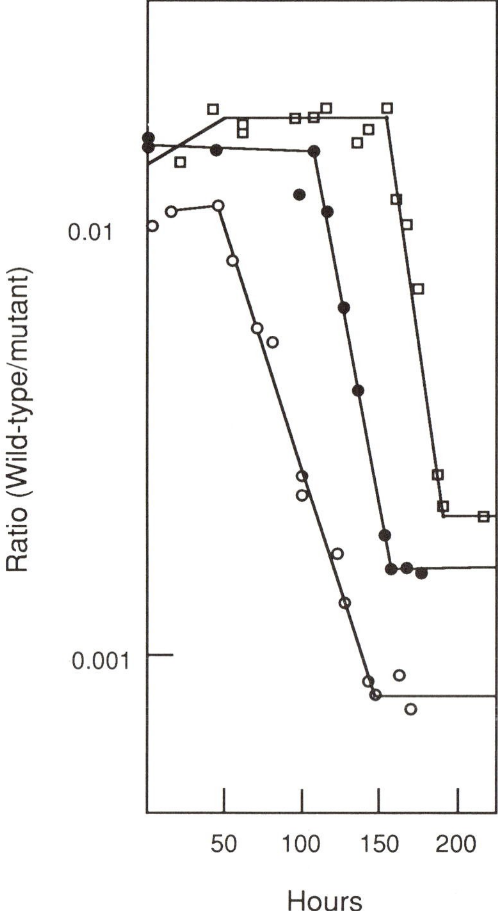

Figure 7.1. Three experiments showing selection against wildtype *E. coli* in favor of tryptophan auxotrophs. Bacteria were grown in chemostats enriched for tryptophan. The limit of sensitivity of the chemostats were approximately 0.5% growth rate differential. Such a differential may be insufficient to detect fitness differences between naturally occurring allozyme variants (see text). (From Dykhuizen 1978.)

occurring allozymes of 6-phosphogluconate dehydrogenase (Dykhuizen and Hartl 1980). In the normal *E. coli* K12 genetic background, all four alleles were found to be selectively neutral. But the limit of detection of selective differences between strains was only 0.5%. How does this finding relate to the problem of whether the allozymes are selectively neutral in nature? Practically speaking, it bears very little relationship. To see this, consider a simple theoretical result from population genetics: for a mutation to have no selective effect it must have an apparent selective disadvantage (or advantage) no greater than the reciprocal of population size. This means that even if the population size of *E. coli* was only 10^6 (the same size as *Drosophila*!), estimated selection differences as small as 0.000001 could be considered nonneutral. Compare this value to the value for the limit of their detection system, 0.005. No amount of experimental manipulation will allow us to know whether apparently neutral allozymes of *E. coli* are in fact selectively neutral in nature if their fitness effects are sufficiently small.

I will now consider two kinds of epistemological limits to knowing in evolutionary biology: when different evolutionary forces produce overlapping or indistinguishable patterns of change, and when multiple evolutionary forces cannot be separated from one another. Both kinds of occurrences are common in molecular evolutionary biology. Together they define a practical limit to reductionism.

An example of the first type—when more than one force produces the same pattern of change—can be taken from my own work on distinguishing the forces governing polymorphism. We have recently developed a method for detecting when a region of DNA containing a polymorphism is being positively maintained by natural selection (Hudson et al. 1987; Kreitman and Hudson 1991). Taking advantage of a theoretical result that shows that neutral mutations in close proximity to a site under balancing selection are expected to accumulate to higher-than-neutral levels (Watterson 1975), we devised a way to find regions of DNA having exceptionally high levels of silent variation. Such a region must have one or more polymorphisms being positively acted on by selection. An example of such a pattern is shown in figure 7.2 for alcohol dehydrogenase in *D. melanogaster*. The question is, Can we know, from this analysis, what kind of selection is acting? The answer is no. Denniston and Crow (1990) have recently shown that it is formally impossible to distinguish between overdominant selection and frequency-dependent selection; both lead to exactly the same equation for allele frequency change. No amount of nucleotide data can distinguish between the two kinds of selection.

It is probably naive to think that at any particular point in time only one evolutionary force dominates the evolutionary process in its contribution to the patterning of polymorphism. Certainly there is generally

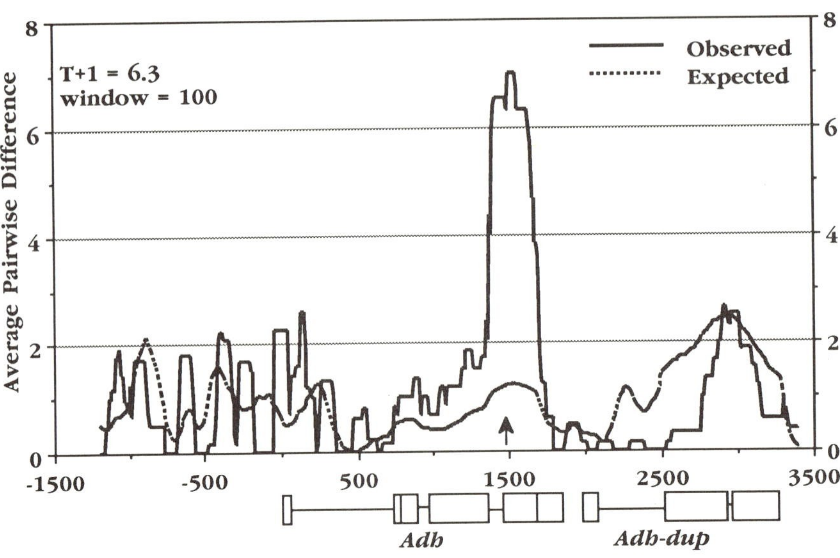

Figure 7.2. Sliding window for *Adh^f/Adh^s* allozyme comparison in eleven *D. melanogaster* alleles. The observed values represent the average pairwise number of nucleotide differences between *Adh^f* and *Adh^s* alleles in a window of one hundred silent sites. The graph provides a visual representation of the patterns of silent nucleotide polymorphism in a two-gene region. The expected values were calculated from between-species data (*D. melanogaster* and *D. simulans*). Under the neutral theory of molecular evolution the two curves should be the same. Instead, there is a statistically significant excess of silent polymorphism centered around the allozyme polymorphism (arrow at position 1490). The excess may reflect the presence of a balanced polymorphism of *Adh^f* and *Adh^s*. (From Kreitman and Hudson 1991.)

more than one evolutionary force acting on a population at any point in time, and the same may apply to individual genes. Genetic drift is always operating in all populations and ineluctably leads to interpopulational genetic differentiation; natural selection also produces local adaptation and divergence (although it can also maintain genetic uniformity); migration opposes differentiation; founder events and other historical factors also shape patterns of variation between populations. Local differentiation may arise by any one of these mechanisms but most often it is the sum of all of them, and no one has yet been able to disentangle all of these causes. Although DNA-based measurement of polymorphism allows us to quantify the differences between populations with very great precision, we still do not know how to distinguish among the multiple causes of local differentiation.

Molecular tools allow us to investigate the genetic material at the highest possible level of resolution—the nucleotide sequence. But reducing a gene to its constituent bases is not the same as reducing a problem to a set of strong alternatives. This is especially true for evolutionary questions. The precision of measurement has lulled many evolutionists into a false sense of expectation for reductionism and this, in turn, has led to unrealistic hopes for solving evolutionary questions with molecular approaches. In evolution, input variables are set by nature, not the experimenter. With polymorphism data, for example, certain variables either are difficult to estimate with great precision (heterozygosity per nucleotide site) or are effectively unmeasurable (the recombination parameter). And if many evolutionary forces are acting on a gene at the same time, it may be impossible for them to be disentangled. Molecular approaches will continue to make important contributions to evolutionary biology, but they will not do for evolution what they have already accomplished for genetics and development.

Financial/Political Limitations

Not the least of the problems facing ecology and evolution in its use of molecular tools and approaches is the tremendous differential in access to resources between the two sciences. Incredibly, the United States spends about 2200% more on health-related research than on all the environmental sciences combined. The annual research budget for molecular biology at Princeton University alone, for example, currently around $10 million, substantially exceeds the U.S. National Science Foundation budget for all of ecology and systematics. Given the alarming decline in environmental conditions and the obvious human-health implications, there is absolutely no logical explanation for such a discrepancy. If ecology and evolution are to compete within biology for the attention of the best young minds, the field must have better resources. If for no other reason, this financial wedge (we all know that universities favor departments doing overhead-generating research) will continue to thwart the development of molecular approaches to the study of evolution.

Prospects for the 1990s

In spite of all its limitations, the tools and discoveries of molecular biology are providing evolutionists with extraordinary new opportunities.

I have already touched on those in population genetic research and in systematics. Ecology is now also benefiting from molecular research. Just as the introduction of allozymes twenty-five years ago brought with it a frenzied attempt to unite ecology with genetics, molecular genetics is now reinvigorating attempts at this union. The novel applications of highly variable loci to determine the maternity or paternity of individuals within natural populations—for organisms as diverse as mice and men, birds and trees—exemplify the breadth of this effort.

Although John Bonner's hopes for uniting the two biologies, in the broad sense, may not be possible or at least will not be consummated through molecular biology, the next ten years will still afford an opportunity to begin the job of understanding morphology and development at the molecular, genetical, and evolutionary levels. So, to complete my discussion of the intellectual meeting grounds for the two biologies, I will describe my own hopes for uniting patterns of evolutionary change with the processes that drive the change. Three advances in molecular biology and genetics and one advance in molecular population genetics are making this union possible: rapid improvements in genetic mapping of complex traits; the likelihood of complete genome sequences; the genetic dissection of morphology and development; and the ability to detect positive selection in molecular evolution.

Already in *Drosophila* (and also in *Caenorhabditis*) the genetics of early embryonic development is remarkably well understood, and the genetics of development of certain complex traits in the adult fly—the compound eye and bristle pattern formation, to name two—is rapidly advancing. Evolutionists can look forward in the near future to having available the developmental genetics of many interesting morphological and developmental traits.

With this knowledge, two kinds of questions will, in principle, be answerable. First, we will be able to explore the repertoire of genetic causes underlying phenotypic changes in evolution. Whether or not multiple genetic pathways can produce similar phenotypic change remains largely a mystery. For example, hybrid sterility is practically a universal feature associated with speciation. This sterility is not always symmetrical with respect to the sexes. According to Haldane's rule, when only one sex is sterile it is the heterogametic sex. Furthermore, the sterility is often caused by factors located on the sex chromosome (Coyne and Orr 1989). Is there a common genetic mechanism to explain the origin of hybrid sterility, the curious asymmetry of its occurrence between the sexes, and the involvement of the sex chromosome? It seems likely to me that this question will be answered with molecular genetic approaches as particular instances of sterility genes are mapped, cloned, and compared among species.

Second, we will be able to ask whether particular phenotypic changes in evolution have been driven by natural selection. To do this requires another kind of advance, this one coming from molecular population genetics. One of the wonderful features of DNA comparison is that it can be made at many different time points along the evolutionary process. Of course, with molecular evolution we are projecting back in time rather than forward—from current populations back to their common ancestors. So, for example, it is possible to quantify the amount of nucleotide polymorphism within a species and compare this value to the amount of nucleotide divergence between species for the same DNA region. With sufficient data for two time points in the evolutionary process, it is now possible to model whether certain evolutionary forces, such as genetic drift and positive selection, or balancing selection is compatible with the observed patterns of variation.

Using this approach, as shown in figure 7.3, we have recently been able to show that amino acid substitutions in alcohol dehydrogenase protein evolution in certain *Drosophila* species occurs at a rate that is several-fold greater than can be accounted for by genetic drift alone (McDonald and Kreitman 1991). The amino acid substitutions of *Adh* in these species may have been driven by adaptive evolution. Therefore, by comparing patterns of DNA sequence variation and applying population genetic theory we can now connect patterns of evolutionary change with the process of change. As genes that are required for the correct development and expression of particular morphologies are identified, the same kind of analysis may lead us to the evolutionary mechanisms for phenotypic change.

Evolutionary biology should not shy away from its own questions and research traditions. Neither should it ignore developments in molecular biology. We will, no doubt, learn a great deal more about evolution from molecular biology in the 1990s, even though, in answer to the question posed in the title, molecular biology will not solve evolution. Like counting to infinity, we will edge ever closer to fulfilling John Bonner's aspirations for a united biology without ever quite getting there.

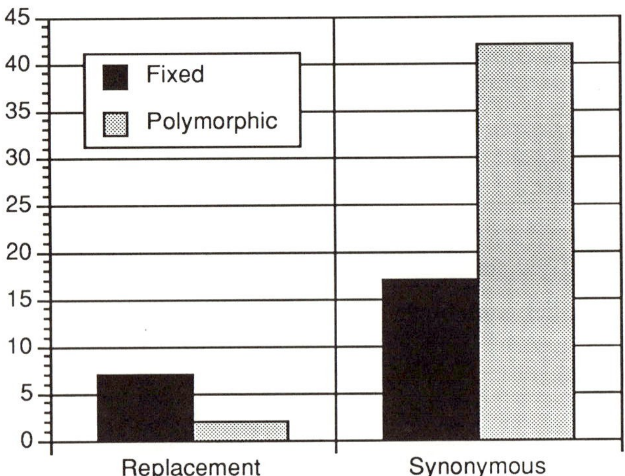

Figure 7.3. Number of amino acid replacement and synonymous substitutions in alcohol dehydrogenase in three species of *Drosophila*. Fixed differences occur between species exclusively; polymorphic differences occur within species. Under the neutral theory of molecular evolution the ratio of fixed : polymorphic substitutions is expected to be the same for synonymous and replacement changes. A G-test of independence can be used to test the null hypothesis (G = 7.43, P = 0.006). The departure from neutrality is in the direction of having a greater than expected number of fixed amino acid replacement changes between species. These changes may have resulted from the fixation of adaptively favored mutations. (From McDonald and Kreitman 1991.)

Summary

John Bonner's research has exemplified how problems in developmental biology could and should be illuminated by evolutionary considerations and, conversely, how the workings of evolution can be elaborated from considerations of development. Today, almost all suborganismal processes, such as pattern formation in development or cell-cell communication, are studied using a single set of tools, largely developed over the last two decades, which altogether define the field of molecular biology. This same set of tools has been enthusiastically adopted by evolutionary biologists, but with the express purpose of helping to understand processes above the level of the individual—evolutionary mechanisms and phylogeny. The revolutionary impact molecular biology is having on medicine, genetics, and development, coupled with the explosion of molecular applications to evolutionary biology, has led to a general euphoria about the prospects for solving many fundamental but recalcitrant problems in evolution. My purpose in this chapter is to explore some of the assumptions behind this euphoria. I argued that the success of molecular biology is also the success of reductionism, but that this same reductionism may not always be achieved in dissecting evolutionary problems. There are certain limitations to knowledge in evolutionary reconstruction, which not even the most precise measurement can overcome. Together, they define a practical limit to the potential of molecular approaches for solving evolutionary problems.

8

Overview: Variation and Change

PETER R. GRANT

Evolutionary biology is the study of processes and products of evolution. Its practitioners try to understand two classes of phenomena: (1) the diversity of organisms at all taxonomic levels; and (2) the nature of traits or characters and their variation, extending from the size, composition, and function of the genome to the size, composition, and functions of adults, as well as all the stages in between by which they are connected. Thus evolutionary biology is concerned with explaining the properties of individuals and the properties of taxa.

Contributors to this book were chosen to span the total broad range of evolutionary biology by representing six different areas. As the chapters demonstrate there are clear connections between adjacent areas. For example, connections are made explicit between molecular and cellular processes (Kreitman, Kirschner), between cellular and developmental processes (Kirschner, Buss and Dick) and between developmental and behavioral processes (West-Eberhard). My first purpose in this closing chapter is to reemphasize these connections with a less obvious illustration: how the development of evolutionary thinking in cellular biology can help us to understand a problem of explosive diversification that occurred more than five hundred million years ago. It is also an example of the connection between taxonomic diversity and trait variation.

My second purpose is to highlight two problem areas or major themes in current evolutionary biology. They emerged, without any prompting from the editors, in the independent contributions to this book. The first is to understand how major changes in evolution occurred. By this we normally mean major reorganization of the body plan of organisms. The problem is more one of elucidating mechanisms intrinsic to the organisms—genetic, cellular, and developmental—than those forces arising from the external environment. The second is to understand variation at all levels of complexity. These subjects—variation and major changes—are interrelated and complex. They will be exercising the minds of evolutionary biologists well into the next century.

Major Changes

Given enough time, small changes will add up to large ones. Even though rates of evolutionary change are constrained by genetic and developmental mechanisms, the large amount of evolutionary change that must have occurred to give rise to differences between major taxa, such as annelids and arthropods, poses no special problem, providing that time has been sufficient. A problem does arise when macroevolution has to be reconciled with a short time interval. The opening chapter by Valentine discusses the most celebrated examples of this.

According to the fossil record the Vendian (late Precambrian) and early Cambrian periods witnessed an extraordinary explosive radiation of metazoans at ordinal, class, and phylum levels in a comparatively short time, on the order of ten million years, perhaps even less. The phenomenon lacks precise, quantitative definition, yet seems undeniable. Certainly there were more invertebrate body plans at that time than now (Valentine 1969), and the numbers are probably underestimated as many of the early species defy classification (Briggs and Conway Morris 1986). Remarkably, almost half of the 155 known orders of that time, distributed among an unknown number of phyla, do not belong to any of the phyla living today (Valentine). Why and how did so many, fundamentally different, body plan architectures arise? Conway Morris (1985) refers to this as a period of metazoan experimentation, while Buss and Dick more specifically suggest it is experimentation with segmental design (see also Jacobs 1990).

Valentine's solution to the problem, borrowed from John Bonner's ideas about the generation of complexity (Bonner 1988), invokes the key notion of gradual evolution of different cell types, which set the stage for subsequent radiation by providing the potential for organismic diversity. This in turn was founded upon the establishment of a hierarchy of pattern formation genes, modifiable through mutation. Thus some major changes in genetic architecture must have been occurring, according to this view, to allow the phenotypic diversification to unfold.

Kirschner's chapter, surveying the cellular and developmental mechanisms responsible for major evolutionary changes in the structure of organisms in general, provides pertinent, broadly based insights into why cell types differ. Using present-day organisms from very different taxa, he finds extreme conservatism in structural and regulatory genes governing eurkaryote cells. Against this background of relative uniformity, a limited but crucial input of novel gene products facilitated some major changes, such as the evolution of reptilian scales and avian feathers. However, major changes appear to have been the result more often of fundamental cellular processes being recombined for new functions. Presumably genes were involved in the recombination, as well as in the regulation of the cellular processes themselves.

This all adds up to a picture of small genetic changes having had profound effects upon developmental mechanisms, interactions, and their products when they caused small rearrangements in cellular processes. The rearrangements may have had little immediate consequence; it is what they facilitated that was profound. The history of experimentation with segmental design was thus a history of experimentation with combinations of cellular processes. Viewed in this light the Precambrian-Cambrian explosion and other rapid radiations appear less enigmatic because the amount of constitutive genetic change required appears to have been small.

Phenotypic plasticity in adult form can be a possible route to evolutionary novelty, and is relevant to this issue as well. Geneticists are well aware that environmental effects which modify phenotypes before the action of natural selection are very important for an understanding of evolutionary change (e.g, Scharloo 1984). The modifications themselves will often be adaptive, though this need not necessarily be so. Thus mammalian bone, muscle, and other connective tissues are capable of adaptively responding to changes in loading during use. Skeletal material becomes remodeled when placed under strain (Biewener 1990). The adult form of various intertidal invertebrates is conditioned by physical forces of water movement during growth, and plants are well known to be plastic in growth form according to the conditions they experience. The same flexibility and environment sensitivity is shown at the level of individual cells in organisms, in the proteins that regulate the cell cycle and in the microtubules involved in mitosis (Kirschner).

West-Eberhard has combined this notion of condition or environment sensitivity with the operation of developmental switches to produce a scheme for generating alternative phenotypic states. Even though they are not different genotypes, either can lead to evolutionary change, which means that the potential for change is enhanced by the coexistence of two or more alternative phenotypes in the population. It would be worth exploring the possible genetic variation underlying the way the switches are sensitive to the environment, either by applying the techniques of quantitative genetics, as in the analysis of diapause in insects (Dingle et al. 1977) and copepods (Hairston and Dillon 1990), or with the same approaches that have yielded fascinating insights into genetic variation at the molecular level (Kreitman).

Kirschner's conclusion, that the study of development offers the best opportunity for understanding the mechanisms of evolution, is echoed by others in this book. Certainly the dissection of the genetic basis of development in organisms such as *Caenorhabditis* (Wood 1988) and *Drosophila* (Bate and Martinez-Arias 1992) has been one of the most illuminating areas of evolutionary biology in recent years. Development is more than a means to an end. Even when agents of natural selection act solely on phenotypic variation among adults they are, in essence, distinguishing among various developmental patterns, however small the differences. And when those patterns vary genetically, selection on adults results in the evolution of development. Kirschner's conclusion is broad.

Variation

G. Evelyn Hutchinson dared to ask, "Why are there so many kinds of animals?" As supervisor of my postdoctoral research, he was my last formal teacher. All of my previous teachers would have labeled his bold question metaphysical, to be addressed solely by "pure speculation." Hutchinson (1959) challenged us to make this a scientifically answerable question, and sketched a map to help in the search for answers. Actually there are two questions wrapped up in one: Why are there so many species (a question of numbers), and why do they differ as much as they do (a question of variety)?

Hutchinson's map identified several salient factors linked to the supposition that a complex trophic organization of a community is more stable than a simple one, and that there is a natural progression from instability to stability, hence several species will accumulate in a community through time. Limits on the processes are set by the tendency for food chains to shorten or become blurred, by unfavorable physical factors, by the fineness of possible subdivisions of niches, and by those characters of the environmental mosaic that permit a greater diversity of small than of large allied species. It is an ecological map.

The challenge was picked up by MacArthur (1970, 1972), May (1973), and others, and the map continues to be redesigned, most notably by incorporating a greater variety of biotic interactions, in both their primary and second-order effects, and by being explicit about scale. However, the question is partly evolutionary (Hutchinson 1968), yet evolutionary considerations have been given minor attention. This is not to deny the active investigation of speciation as a means of generating diversity. It is in fact an area of much current interest (Otte and Endler 1989). But the end of many of these inquiries is a hypothesis or two for how two species were formed from one. This is a small, albeit crucial, ingredient of the answer, addressing the numerical issue of the diversity question more than the variety issue. There is a need to treat the question of why there are so many kinds of all species, not just animals, as a problem in evolutionary ecology. As with attempts to understand sexual reproduction and complex life cycles, the problem is to elucidate the ecological and evolutionary factors responsible for their origin, on the one hand, and for their maintenance, on the other. The chapter by Bell in this volume and companion papers (Bell 1990a, 1990b) build on the groundwork laid by Levene (1953), Maynard Smith (1966), and others by exploring, as a beginning, the influence of environmental (ecological) variation on genetic variation within single species.

Accounting for species diversity is a daunting challenge. But then why restrict the question of diversity to species? Why are there so many kinds of organisms that we group into genera, families, orders, and so on, and why do they differ among themselves as much as they do? This is perhaps more an evolutionary problem than an ecological one. There are theories to explain patterns of local variation in number of species, such as on islands (MacArthur and Wilson 1967; Williamson 1981), but there is no general theory to account for patterns of diversity at any taxonomic level. Perhaps we should not expect there to be a single general theory if evolutionary processes such as speciation differ substantially among phyla or kingdoms—between plants and animals, for example.

Two small steps necessary for the development of theory have been taken recently by paleontologists. The first has been to tackle with the aid of computers the monumental task of quantitatively describing the phenomena to be explained, that is, the fluctuations through geological time of the sizes and compositions of various taxa. The second has been to use the correlative approach in order to explain, in terms of known biotic and abiotic factors, the waxing and waning of individual taxa, such as the orders of fishes in the Paleozoic, or the families of Ammonoids throughout their history (e.g., see Valentine 1985; Vermeij 1987).

Just as one can look upwards in the organic hierarchy from the level of species and pose the same question about diversity at higher taxonomic levels, so can one look downwards and ask the question of variation at the level of populations within species, at the level of individuals within populations, at the level of stages in the lives of individual organisms, and at the level of constituent parts within organisms. In this volume Buss and Dick remind us of the challenge to answer the third question. Some life cycles are extraordinarily complex. Buss and Dick argue that, far from being aberrant features of a few invertebrates, complex life cycles should occupy a central place in evolutionary biology as they sharply focus our attention on the level(s) at which natural selection is supposed to operate.

Explaining variation among individuals in a population has largely been a genetical problem, stimulated by the discovery more than twenty five years ago of large amounts of electrophoretically detectable genetic variation in natural populations. While much if not most of the variation at the molelcular level is effectively neutral, at least some is selectively maintained (Kreitman; see also McDonald and Kreitman 1991). At the level of traits, variation is enhanced by mutation and recombination, also by selection under some circumstances. It is depleted by most forms of selection and by drift. Insofar as selective forces originate in the external environment, explaining variation is at least partly a problem for ecologists to solve (Grant and Price 1981). Thus Bell's chapter is pivotal in discussing the ecological genetics of variation within species, as well as in its relevance to questions of species diversity.

Extraordinary advances have been made in the last twenty-five years in our understanding of how novel genetic variation arises. Unsuspected mechanisms have been revealed, poorly known ones have been demonstrated, and genomes are in a state of flux far more than was ever thought possible when the Modern Synthesis was born. There are genes that enhance mutation at other loci, genes that enhance recombination, genes that are found only in zones of hybridization and genes that move around the genome. It is at this point of considering how levels of genetic variation may be increased that we return to the previous theme and establish a link with the origin of major changes. Part of the answer to the riddle of the rapid radiations may be that high rates of evolution were facilitated by the generation of unusual levels of genetic variation. We can suggest plausible genetic mechanisms for how this might be, but these will always be speculative because the species, at least those in the Vendian explosion, have long since disappeared. Can we suggest plausible environmental reasons as well?

To answer this question, an obvious starting point is the environmental record to be read from the rocks of the late Precambrian. The record has recently been interpreted to reveal a series of ice ages, the incipient breakup of a super continent and strong fluctuations in organic carbon and oxygen (Knoll and Walker 1990). Environmental change at this time appears to have been profound and rapid. Modern populations are stressed when the environment changes. Parsons (1988) has summarized laboratory data showing that recombination is increased under nutritional, thermal, and behavioral stress in *Drosophila* and mice. Thus an argument can be made for supposing that levels of genetic variation were enhanced in the late Precambrian, and rates of evolution increased, as a result of stresses caused by rapid changes in the environment. These rapid evolutionary changes should be viewed in the context of a world relatively depauperate biotically, in which ecological opportunities for diversification were plentiful (Valentine 1975; Erwin et al. 1987). Opportunities were constantly being created by environmental change directly, as well as indirectly through presumably frequent extinctions, which is one possible reason why diversity remained low at the level of species while it increased at ordinal and higher levels.

Recent studies have also identified hybridization as a potentially important process in the generation of diversity. Genetic novelties have been found restricted to hybrid zones of mammals, birds, reptiles, amphibia, molluscs, and insects. They occur in zones of intergradation between subspecies as well as in true hybrid zones between species. They may have arisen in the zone of hybridization itself through the breakdown of suppression systems and the release of mutator activity, through transposon-induced hybrid dysgenesis, or through intragenic recombination (Woodruff 1989). If genetic exchange between taxa was more prevalent in the early days of metazoan evolution, it may have contributed to the rapid diversification of those taxa. The place to look for hybridization is the Burgess Shale fauna, since samples of some of the genera are large (Conway Morris 1986) and phenotypic measurements are possible.

Kreitman cautions that molecular biology and evolutionary biology are largely separate disciplines, with separate traditions and separate goals, and warns that not too much should be expected from a mutualistic association between them. While this serves a useful purpose, I nevertheless believe that the problem of understanding variation, and of understanding how that variation is transformed in major ways, is going to be solved through evolutionary studies in which molecular genetics will play a conspicuous and essential role.

Conclusions

To reiterate, this volume pinpoints and highlights exciting developments in six areas of evolutionary biology, and some of the connections between them. It does not attempt to be comprehensive. There is little or nothing here on some currently challenging topics, including human evolution, the reconstruction of phylogeny, the basis of sexual selection, the origin and maintenance of sex, molecular drive and transposable elements, neuro-endocrinological control and regulatory systems, and—Darwin's mystery of mysteries—the origin of life itself. Evolutionary biology has both a rich history and a rich future.

References

Akam, M. 1987. The molecular basis for metameric pattern in the Drosophila embryo. *Development* 101: 1–22.

Akam, M., I. Dawson, and G. Tear. 1988. Homeotic genes and the control of segment diversity. *Development* 104 (Suppl.): 123–133.

Alberts, B., D. Bray, J. Lewis, M. Raff, K. Roberts, and J. D. Watson. 1989. *Molecular Biology of the Cell*. New York and London: Garland Publishing.

Antonarakis, S. E., P. Oettgen, A. Chakravarti, S. L. Halloran, R. R. Hudson, L. Feisee, and S. K. Karathanasis. 1988. DNA polymorphism haplotypes of the human apolipoprotein APOA1-APOC3-APOA4 gene cluster. *Human Genetics* 80: 265–273.

Aquadro, C. F. 1989. Contrasting levels of DNA sequence variation in *Drosophila* species revealed by ''six-cutter'' restriction map surveys. In M. Clegg and S. O'Brien, eds., *Molecular Evolution,* UCLA Symposium on Molecular and Cellular Biology, New Series, vol. 122. New York: Alan R. Liss, Inc.

Armstrong, D. P. 1984. Why don't cellular slime molds cheat? *J. Theor. Biol.* 109: 271–283.

Baker, G. A., M. R. Huberty, and F. J. Veihmeyer. 1950. An eleven-year uniformity trial of an irrigated barley field. *Agronomical Journal* 42: 267–270.

Ballard, W. W. 1981. Morphogenetic movements and fate maps of vertebrates. *American Zoologist* 21: 391–399.

Barrett, J. A. 1981. The evolutionary consequences of monoculture. Pp. 209–248 In J. A. Bishop and L. M. Cook, eds., *Genetic Consequences of Man-made Change*. London: Academic Press.

Bate, M., and A. Martinez-Arias, eds. 1992. *The Development of* Drosophila. New York: Cold Spring Harbor Press.

Bell, G. 1982. *The Masterpiece of Nature*. London: Croom-Helm, and Berkeley: University of California Press.

Bell, G. 1990a. The ecology and genetics of fitness in *Chlamydomonas:* I. Genotype-by-environment interaction among pure strains. *Proc. Roy. Soc. London* B240: 295–321.

Bell, G. 1990b. The ecology and genetics of fitness in *Chlamydomonas:* II. The properties of mixtures of strains. *Proc. Roy. Soc. London* B240: 323–350.

Bell, G. 1991a. The ecology and genetics of fitness in crop plants. Unpublished manuscript.

Bell, G. 1991b. The ecology and genetics of fitness in *Chlamydomonas:* III. Genotype-by-environment interaction within strains. *Evolution* (in press).

Bell, G. 1991c. The ecology and genetics of fitness in *Chlamydomonas:* IV. The properties of mixtures of genotypes of the same species. *Evolution*. 45:1036–1046.

Bell, G. 1991d. The ecology and genetics of fitness in *Chlamydomonas:* V. The relationship between genetic correlation and environmental variance. Unpublished manuscript.

Bell, G., and M. Lechowicz. 1991. The ecology and genetics of fitness in forest plants: I. Environmental heterogeneity measured by explant trials. *Journal of Ecology* (in press).

Bell, G., and J. Maynard Smith. 1987. Short-term selection for recombination among mutually antagonistic species. *Nature* 327: 66–68.

Bell, G., M. Lechowicz, and D. Schoen. 1991. The ecology and genetics of fitness in forest plants: III. Environmental variance in native populations of *Impatiens pallida*. *Journal of Ecology* (in press).

Bennett, N., and A. Sitaramayya. 1988. Inactivation of photoexcited rhodopsin in retinal rods: The roles of rhodopsin kinase and 48-kDa protein (arrestin). *Biochemistry* 27: 1710–1715.

Bergstrom, J. 1990. Precambrian trace fossils and the rise of bilaterian animals. *Ichnos* 1: 3–13.

Berrill, N. J. 1931. Regeneration in *Sabella pavonina* (Sav.) and other sabellid worms. *J. Exp. Zool.* 58: 495–523.

Berrill, N. J. 1952. Regeneration and budding in worms. *Biol. Rev.* 27: 401–438.

Bieber, A. J., P. M. Snow, M. Hortsch, N. H. Patel, J. R. Jacobs, Z. R. Tracquina, J. Schilling, and C. S. Goodman. 1989. Drosophila neuroglean: A member of the immunoglobulin superfamily with extensive homology to the vertebrate neural adhesion molecule L1. *Cell* 59: 447–460.

Biewener, A. A. 1990. Biomechanics of mammalian terrestrial locomotion. *Science* 250: 1097–1104.

Biggin, M. D., and R. Tjian. 1989. Transcription factors and the control of Drosophila development. *Trends in Genetics* 5: 377–383.

Bishop, J. M. 1987. The molecular genetics of cancer. *Science* 235: 305–311.

Bloomquist, B. J., R. D. Shortridge, S. Schneuwky, M. Perdeu, C. Montell, H. Steller, G. Rubin, and W. L. Pak. 1988. Isolation of a putative phospholipase C gene of Drosophila, nor pA, and its role in phototransduction. *Cell* 54: 723–733.

Bock, W. J. 1979. The synthetic explanation of macroevolutionary change—a reductionist approach. *Bull. Carnegie Mus. Nat. Hist.* 13: 20–69.

Bogert, A. 1894. *Die Thaliacea der Plankton-Expedition. Vertheilung der Doliolen. Ergebnisse der Plankton-Expedition der Humboldt-Stiftung, v. 2E.* Kiel: C. Lipsius & Tischer.

Bonner, J. T. 1958. *The Evolution of Development*. New York: Cambridge University Press.

Bonner, J. T. 1965. *Size and Cycle*. Cambridge, Mass.: Harvard University Press.

Bonner, J. T. 1977. *On Development*. Cambridge, Mass.: Harvard University Press.

Bonner, J. T. 1980. *The Evolution of Culture in Animals*. Princeton, N.J.: Princeton University Press.

Bonner, J. T., ed. 1982. *Evolution and Development*. Berlin: Springer-Verlag.

Bonner, J. T. 1988. *The Evolution of Complexity by Means of Natural Selection*. Princeton, N.J.: Princetion University Press.

Borst, P., and D. R. Greaves. 1987. Programmed gene rearrangements altering gene expression. *Science* 235: 658–667.

Bourne, H. R. 1988. Do GTPases direct membrane traffic in secretion? *Cell* 53: 669–671.

Brandon, R. N. 1990. *Adaptation and Environment*. Princeton, N.J.: Princeton University Press.

Brasier, M. D. 1989. Towards a biostratigraphy of the earliest skeletal biotas. In J. W. Cowie and M. D. Brasier, eds., *The Precambrian-Cambrian Boundary*, 117–165. Oxford: Clarendon Press.

Briggs, D.E.G., and S. Conway Morris. 1986. Problematica from the Middle Cambrian Burgess Shale of British Columbia. In A. Hoffman and M. H. Nitecki, eds., *Problematic Fossil Taxa*, 167–183. New York: Oxford University Press.

Burt, A., and G. Bell. 1987. Mammalian chiasma frequencies as a test of two theories of recombination. *Nature* 326: 803–805.

Buss, L. W. 1982. Somatic cell parasitism and the evolution of somatic tissue compatibility. *Proc. Nat. Acad. Sci. USA* 79: 5337–5341.

Buss, L. W. 1983. Evolution, development and the units of selection. *Proc. Nat. Acad. Sci. USA* 80: 1387–1391.

Buss, L. W. 1987. *The Evolution of Individuality*. Princeton, N.J.: Princeton University Press.

Byers, P. H. 1989. Inherited disorders of collagen gene structure and expression. *Amer. J. Med. Gen.* 34: 72–80.

Cann, R. L., M. Stoneking, and A. C. Wilson. 1987. Mitochondrial DNA and human evolution. *Nature* 325: 31–35.

Charlesworth, B., R. Lande, and M. Slatkin. 1982. A neo-Darwinian commentary on macroevolution. *Evolution* 36: 474–498.

Chen, J., X. Hou, and H. Lu. 1989. Early Cambrian netted scale-bearing wormlike sea animal. *Acta Palaeontologica Sinica* 28: 1–16.

Clark, R. B. 1964. *Dynamics in Metazoan Evolution*. Oxford: Clarendon Press.

Clark, R. B. 1965. Endocrinology and the reproductive biology of polychaetes. *Oceanogr. Mar. Biol. Ann. Rev.* 3: 211–255.

Clark, R. B. 1966. The integrative action of a worm's brain. *Symp. Soc. Exp. Biol.* 20: 345–379.

Clark, R. B. 1969. Endocrine influences in annelids. *Gen. Comp. Endocrinol.* 2 (Suppl.): 572–581.

Clark, R. B. 1979. Radiation of Metazoa. In M. R. House, ed., *The Origin of Major Invertebrate Groups*, 55–102. London: Academic Press.

Clark, R. B., and P.J.W. Olive. 1973. Recent advances in polychaete endocrinology and reproductive biology. *Oceanogr. Mar. Biol. Ann. Rev.* 11: 175–222.

Conway Morris, S. 1976. *Nectocaris pteryx*, a new organism from the Middle Cambrian Burgess Shale of British Columbia. *Neues Jahrbuch für Geologie und Paläeontologie* 12: 705–713.

Conway Morris, S. 1985. The Middle Cambrian metazoan *Wiwaxia corrugata* (Matthew) from the Burgess Shale and *Ogygopsis* shale, British Columbia, Canada. *Phil. Trans. Roy. Soc. London* B307: 507–582.

Conway Morris, S. 1986. The community structure of the Middle Cambrian phyllopod bed (Burgess Shale). *Palaeontology* 29: 423–467.

Cook, S. A., and M. P. Johnson. 1968. Adaptation to heterogeneous environments:

I. Variation in heterophylly in *Ranunculus flammula* L. *Evolution* 22: 496–516.

Coyne, J. A., and M. Kreitman. 1986. Evolutionary genetics of two sibling species, *Drosophila simulans* and *D. sechellia*. *Evolution* 40: 673–691.

Coyne, J. A., and H. A. Orr. 1989. Two rules of speciation. In D. Otte and J. A. Endler, eds., *Speciation and Its Consequences*. Sunderland, Mass.: Sinauer Associates.

Darnell, J., H. Lodish, and D. Baltimore. 1986. *Molecular Cell Biology*. New York: Scientific American Books.

Darwin, C. 1868 (1896). *The Variation of Animals and Plants under Domestication*. New York: D. Appleton and Company.

Darwin, C. 1871 (1936). *The Descent of Man and Selection in Relation to Sex*. New York: Modern Library.

Dawkins, R. 1983. *The Extended Phenotype*. Oxford: Oxford University Press.

Denniston, C., and J. F. Crow. 1990. Alternative fitness models with the same allele frequency dynamics. *Genetics* 125: 201–205.

Diamond, M. J., J. N. Miner, S. K. Yoshinaga, and K. R. Yamamoto. 1990. Transcription factor interactions: Selectors of positive or negative regulation from a single DNA binding element. *Science* 249: 1266–1272.

Dingle, H., C. K. Brown, and J. P. Hegmann. 1977. The nature of genetic variance influencing photoperiodic diapause in a migrant insect, *Oncopeltus fasciatus*. *American Naturalist* 111: 1047–1059.

Dobzhansky, T. 1937. *Genetics and the Origin of Species*. New York: Columbia University Press.

Dobzhansky, T. 1970. *Genetics of the Evolutionary Process*. New York: Columbia University Press.

Dover, G. A., and D. Tautz. 1986. Conservation and divergence in multigene families: Alternatives to selection and drift. *Phil. Trans. Roy. Soc. London* B: 275–289.

Ducke, A. 1914. Über Phylogenie und Klassifikation der sozialen Vespiden. *Zoologische Jahrbücher, Abt. für Systematik V* 36: 303–330.

Dykhuizen, D. 1978. Selection for tryptophan auxotrophs of *Escherichia coli* in glucose limited chemostats as a test of the energy conservation hypothesis of evolution. *Evolution* 32: 125–150.

Dykhuizen D., and D. L. Hartl. 1980. Selective neutrality of 6PGD allozymes in *E. coli* and the effects of genetic background. *Genetics* 96: 801–817.

Eberhard, W. G. 1979. The function of horns in *Podischnus agenor* (Dynastinae) and other beetles. In M. S. Blum and N. Blum, eds., *Sexual Selection and Reproduction Competition in Insects*, 231–258. New York: Academic Press.

Eberhard, W. G. 1980. Horned beetles. *Scientific American* 242(3): 166–182.

Eberhard, W. G. 1982. Beetle horn dimorphism: Making the best of a bad lot. *American Naturalist* 119: 420–426.

Edgar, B. A., and P. H. O'Farrell. 1990. The three post blastoderm cell cycles of Drosophila embryogenesis are regulated in G2 by string. *Cell* 62: 469–480.

Eibl-Eibesfeldt, I. 1970. *Ethology: The Biology of Behavior*. New York: Holt, Rinehart and Winston.

Eldredge, N., and S. J. Gould. 1972. Punctuated equilibria: An alternative to phyletic gradualism. In T.J.M. Schopf, ed., *Models in Paleobiology,* 82–115. San Francisco: Freeman, Cooper and Company.

Errede, B., M. Company, and C.A.I. Hitchison. 1987. Tyl sequence with enhancer and mating-type dependent regulatory activities. *Molec. and Cell Biol.* 1: 258–265.

Erwin, D. H., J. W. Valentine, and J. J. Sepkoski, Jr. 1987. A comparative study of diversification events: The early Paleozoic versus the Mesozoic. *Evolution* 41: 1177–1186.

Evans, H. E. 1966. The comparative ethology and evolution of the sand wasps. Cambridge, Mass.: Harvard University Press.

Evans, H. E., and R. W. Matthews. 1975. The sand wasps of Australia. *Scientific American* 233(6): 108–115.

Evans, T., E. T. Rosenthal, J. Youngblau, P. Distal, and T. Hunt. 1983. Cyclin: A protein specified by mRNA in sea urchin eggs that is destroyed at each cleavage division. *Cell* 33: 389–396.

Ewens, W. J. 1979. *Mathematical Population Genetics.* New York: Springer-Verlag.

Federoff, N. V. 1989. About maize transposable elements and development. *Cell* 56: 181–191.

Fedonkin, M. A. 1985a. Systematic description of Vendian Metazoa. In B. S. Sokolov and A. B. Ivanovich, eds., *The Vendian System*, vol. 1, 70–106. Moscow: Nauka (in Russian).

Fedonkin, M. A. 1985b. Paleoichnology of Vendian Metazoa. In B. S. Sokolov and A. B. Ivanovich, eds., *The Vendian System,* vol. 1, 112–117. Moscow: Nauka (in Russian).

Felsenstein, J. 1982. Numerical methods for inferring evolutionary trees. *Quart. Rev. Biol.* 57: 379–404.

Felsenstein, J. 1988. Phylogenies from molecular sequences: Inference and reliability. *Ann. Rev. Gen.* 22: 521–565.

Field, K. G., G. J. Olsen, D. J. Lane, S. J. Giovannoni, M. T. Ghiselin, E. C. Raff, N. R. Pace, and R. A. Raff. 1988. Molecular phylogeny of the animal kingdom. *Science* 239: 748–753.

Filosa, M. F. 1962. Heterocytosis in cellular slime molds. *American Naturalist* 91: 321–325.

Fisher, R. A. 1930. *The Genetical Theory of Natural Selection.* New York: Dover.

Flanders, S. 1969. Social aspects of facultative gravidity and agravidity in Hymenoptera. *Proc. VI Congress Intern. Union Study of Social Insects,* 47–53. Bern.

Ford, E. B. 1961. The theory of genetic polymorphism. In J. S. Kennedy, ed., *Insect Polymorphism,* 11–19.

Franke, H.-D., and H.-D. Pfannenstiel. 1984. Some aspects of endocrine control of polychaete reproduction. *Fortschrit. Zool.* 29: 53–72.

Futuyma, D., and M. Slatkin. 1983. Coevolution. Sunderland, Mass.: Sinauer Associates.

Gadgil, M. 1972. Male dimorphism as a consequence of sexual selection. *American Naturalist* 106: 574–80.

Galton, F. 1889. *Natural Inheritance.* London: Macmillan.

Gans, C. 1989. Stages in the origin of vertebrates: Analysis by means of scenarios. *Biol. Rev.* 64: 221–268.

Gans, C., and R. G. Northcutt. 1983. Neural crest and the origin of vertebrates: A new head. *Science* 220: 268–274.

Gasaway, W. C. 1976. Seasonal variation in diet, volatile fatty acid production and size of the cecum of Rock Ptarmigan. *Comp. Biochem. and Physiol.* 53A: 109–114.

Gehring, W. J., M. Müller, M. Affolter, A. Percival-Smith, M. Billeter, Y. Q. Qian, G. Otting and K. Wüthrich. 1990. The structure of the homeodomain and its functional implications. *Trends in Genetics* 6: 323–328.

Ghiselin, M. T. 1974. *The Economy of Nature and the Evolution of Sex*. London: University of California Press, Ltd.

Ghiselin, M. T. 1988. The origin of molluscs in the light of molecular evidence. *Oxford Surveys Evol. Biol.* 5: 66–95.

Gilinsky, N. L., and R. K. Bambach. 1987. Asymmetrical patterns of origination and extinction in higher taxa. *Paleobiology* 13: 427–445.

Glaessner, M. F. 1976. A new genus of late Precambrian polychaete worms from South Australia. *Trans. Roy. Soc. South Australia* 100: 169–170.

Glaessner, M. F. 1984. *The Dawn of Animal Life*. Cambridge, England: Cambridge University Press.

Goldschmidt, R. B. 1940. *The Material Basis of Evolution*. New Haven, Conn.: Yale University Press.

Gould, S. J. 1977. The return of hopeful monsters. *Natural History* (July): 22–30.

Gould, S. J. 1977. *Ontogeny and Phylogeny*. Cambridge, Mass.: Harvard University Press.

Gould, S. J. 1989. *Wonderful Life*. New York: Norton.

Graham, J. B., R. H. Rosenblatt, and C. Gans. 1978. Vertebrate air breathing arose in fresh waters and not in the oceans. *Evolution* 32: 459–463.

Grant, P. R. 1986. *Ecology and Evolution of Darwin's Finches*. Princeton, N.J.: Princeton University Press.

Grant, P. R., and T. D. Price. 1981. Population variation in continuously varying traits as an ecological genetics problem. *American Zoologist* 21: 795–811.

Guillen, A., J. M. Jailon, J. A. Fehrentz, C. Pantaloni, J. Backaert, and V. Homburger. 1990. A GO-like protein in Drosophila melanogaster and its expression in memory mutants. *EMBO Journal* 9: 1449–1455.

Gyllensten, U. B., S. Jakobssoon, H. Temrin, and A. C. Wilson. 1989. Nucleotide sequence and genomic organization of bird minisatellites. *Nucleic Acids Research* 17: 2203–2214.

Hairston, N. G., Jr., and T. A. Dillon. 1990. Fluctuating selection and response in a population of freshwater copepods. *Evolution* 44: 1796–1805.

Hall, A. 1990. The cellular function of small GTP-binding proteins. *Science* 249: 635–640.

Hamilton, W. D. 1964. The genetical evolution of social behavior, I. and II. *J. Theor. Biol.* 7: 1–52.

Hamilton, W. D. 1980. Sex versus non-sex versus parasite. *Oikos* 35: 282–290.

Hamilton, W. D., R. Axelrod, and R. Tanese. 1990. Sexual reproduction as an adaptation to resist parasites (a review). *Proc. Nat. Acad. Sci. USA* 87: 3566–3573.

Hazel, W.N., R. Smock, and M. D. Johnson. 1990. A polygenic model for the evolution and maintenance of conditional strategies. *Proc. Roy. Soc. London* 242: 181–187.

Henderson, R., and G. F. Schertler. 1990. The structure of bacterio rhodopsin and its relevance to the visual opsins and other seven-hertix G-protein coupled receptors. *Phil. Trans. Roy. Soc. London* B. (Biological Sciences) 326: 379–389.

Herdman, W. A. 1922. Ascidians and Amphioxus. *In* S. F. Harmer and A. E. Shipley, eds., *The Cambridge Natural History,* vol. 7. London: Macmillan.

Herskowitz, I. 1989. A regulatory hierarchy for cell specialization in yeast. *Nature* 342: 749–757.

Hinz, I., D. Kraft, M. Mergl and K. J. Müller. 1990. The problematic *Hadimopanella, Milaculum*, and *Utahphospha* identified as sclerites of Palaeoscolecida. *Lethaia* 23: 217–221.

Hoffman, A. 1989. *Arguments on Evolution: A Paleontologist's Perspective*. New York: Oxford University Press.

Holman, E. W. 1989. Some evolutionary correlates of higher taxa. *Paleobiology* 15: 357–363.

Hudson, R. R., M. Kreitman, and M. Aguadé. 1987. A test of neutral molecular evolution based on nucleotide data. *Genetics* 116: 153–159.

Hull, D. 1980. Individuality and selection. *Ann. Rev. Ecol. Syst.* 11: 311–332.

Hutchinson, G. E. 1959. Homage to Santa Rosalia, or, Why are there so many kinds of animals? *American Naturalist* 93: 145–159.

Hutchinson, G. E. 1965. *The Ecological Theatre and the Evolutionary Play*. New Haven, Conn.: Yale University Press.

Hutchinson, G. E. 1968. When are species necessary? In R. C. Lewontin, ed., *Population Biology and Evolution*, 177–186. Syracuse, N.Y.: Syracuse University Press.

Hutson, V., and R. Law. 1981. Evolution of recombination in populations experiencing frequency-dependent selection with time delay. *Proc. Roy. Soc. London* B213: 345–359.

Hyman, L. H. 1940. *The Invertebrates: Protozoa through Ctenophora*. New York: McGraw-Hill.

Ingham, P. W. 1988. The molecular genetics of embryonic pattern formation in *Drosophila. Nature* 335: 25–34.

Jacobs, D. K. 1990. Selector genes and the Cambrian radiation of Bilateria. *Proc. Nat. Acad. Sci. USA* 87: 4406-4410.

Jaenike, J. 1978. An hypothesis to account for the maintenance of sex within populations. *Evolutionary Theory* 3: 191–194.

Jeffreys, A. J., V. Wilson, and S. L. Thein. 1985. Hypervariable 'minisatellite' region in human DNA. *Nature* 314: 67–73.

Jenkins, R.J.F., C. H. Ford, and G. Gehling. 1983. The Ediacara member of the Rawnsley quartzite: The context of the Ediacara assemblage (Late Precambrian, Flinders Ranges). *J. Geol. Soc. Australia* 30: 101–119.

Jepsen, G. L., E. Mayr, and G. G. Simpson, eds. 1949. *Genetics, Paleontology, and Evolution*. Princeton, N.J.: Princeton University Press.

John, G., and G. Miklos. 1988. *The Eukaryote Genome in Development and Evolution*. London: Allen and Unwin.

Johnson, H. P. 1902. Collateral budding in annelids of the genus *Trypanosyllis*. *American Naturalist* 36: 295–315.

Kelley, S. E., J. Antonovics, and J. Schmitt. 1988. A test of the short-term advantage of sexual reproduction. *Nature* 331: 714–716.

Ketterson, E. D., V. Nolan Jr., L. Wolf, and C. Ziegenfus. Testosterone and avian life histories: The effect of experimentally elevated testosterone on behavior and correlates of fitness in the dark-eyed junco (*Junco hyemalis*). Unpublished manuscript.

Kimmel, C. B. 1989. Genetics and early development of zebrafish. *Trends in Genetics* 5(8): 283–288.

Kimura, M. 1983. *The Neutral Theory of Molecular Evolution*. Cambridge, Mass.: Cambridge University Press.

King, K., H. G. Dohlmann, T. Thorner, M. G. Caron, and R. J. Lefkowitz (1991). Control of yeast mating signal transduction by a mammalian p-2 adrenergies receptor and G_5 alpha subunit. *Science* 250: 121–123.

Kirschner M. W., and T. L. Mitchison. 1986. Beyond self-assembly: From microtubules to morphogenesis. *Cell* 45: 329–342.

Kishimoto, Y. 1986. Phylogenetic development of myelin glycosphingolipids. *Chem. and Physics of Lipids* 42: 117–128.

Knoll, A. H., and J.C.G. Walker. 1990. The environmental context of early metazoan evolution. *Geol. Soc. America Abstracts with Programs* 22 (7): A128.

Korschelt, E., and K. Heider. 1909. *Lehrbuch der Vergleichenden Endwicklungsgeschichte der Wirbellosen Thiere*. Jena: Gustav Fischer.

Krebs, J. R., and N. B. Davies, eds. 1984. *Behavioural Ecology: An Evolutionary Approach*. Oxford: Blackwell Scientific.

Kreitman, M. 1990. Detecting selection at the level of DNA. In R. K. Selander, A. G. Clark, and T. S. Whittam, eds., *Evolution at the Molecular Level*. Sunderland, Mass.: Sinauer Associates.

Kreitman, M., and R. R. Hudson. 1991. Inferring the evolutionary histories of the *Adh* and *Adh-dup* loci in *Drosophila melanogaster* from patterns of polymorphism and divergence. *Genetics* 127: 565–582.

Lake, J. A. 1990. Origin of the Metazoa. *Proc. Nat. Acad. Sci. USA* 87: 763–766.

Lake, J. A. 1991. Tracing origins with molecular sequences: Metazoan and eukaryotic beginnings. *Trends in Biochem. Sci.* 16: 46–50.

Lande, R. 1980. Microevolution in relation to macroevolution. *Paleobiology* 6: 235–238.

Langille, R. M., and B. K. Hall. 1989. Developmental processes, developmental sequences and early vertebrate phylogeny. *Biol. Rev.* 64: 73–91.

Lechowicz, M., and G. Bell. 1991. The ecology and genetics of fitness in forest plants. II. Microspatial heterogeneity of the edaphic environment. *Journal of Ecology* (in press).

Lee, M. G., and P. Nurse. 1987. Complementation used to clone a human homologue of the fission yeast cell cycle control gene *cdc2*. *Nature* 327: 31–35.

Lemke, G., E. Lamar, and J. Patterson. 1988. Isolation and analysis of the gene encoding peripheral myelin protein zero. *Neuron* 1: 73–83.

Levene, H. 1953. Genetic equilibrium when more than one ecological niche is available. *American Naturalist* 87: 131–133.

Levins, R. 1961. Mendelian species as adaptive systems. *General Systems* 6: 31–39.

Levins, R. 1968. *Evolution in Changing Environments: Some Theoretical Explorations*. Princeton, N.J.: Princeton University Press.

Levinton, J. 1988. *Genetics, Paleontology, and Macroevolution*. Cambridge, England: Cambridge University Press.

Li, W.-H., and L. Sadler. 1991. DNA variation in humans and its implications for human evolution. In *Oxford Surveys Evol. Biol.*

Lillie, F. R. 1927. The gene and the ontogenetic process. *Science* 66: 361–368.

Lin, Y. S., M. Carey, M. Ptashne, and M. R. Green. 1990. How different eukaryotic transcription activators can cooperate promiscuously. *Nature* 345: 359–361.

Lindeque, M., and J. D. Skinner. 1982. Fetal androgens and sexual mimicry in spotted hyaenas (*Crocuta crocuta*). *J. Reproduction and Fertility* 67: 405–410.

Lloyd, E. A. 1989. *The Structure and Confirmation of Evolutionary Theory*. Westport, Conn.: Greenwood Press.

Lord, E. M., and J. P. Hill. 1987. Evidence for heterochrony in the evolution of plant form. In R. A. Raff and E. C. Raff, eds., *Development as an Evolutionary Process*, 47–70. New York: Liss.

Lowe, J. B., M. S. Boguski, D. A. Sweetser, N. A. Ecshourbagy, J. M. Taylor, and J. I. Gordon. 1985. Human liver fatty acid binding proteins. *J. Biol. Chem.* 260: 3413–3417.

Lynch, M. 1988. Estimation of relatedness by DNA fingerprinting. *Molec. Biol. and Evol.* 5: 584.

MacArthur, R. H. 1970. Species packing and competitive equilibrium among many species. *Theor. Pop. Biol.* 1: 1–11.

MacArthur, R. H. 1972. *Geographical Ecology*. New York: Harper and Row.

MacArthur, R. H., and E. O. Wilson. 1967. *The Theory of Island Biogeography*. Princeton, N.J.: Princeton University Press.

McDonald, J., and M. Kreitman. 1991. Adaptive protein evolution at the *Adh* locus in *Drosophila*. *Nature* (in press).

Manton, S. M. 1977. *The Arthropods*. Oxford: Clarendon Press.

Margulis, L. 1981. *Symbiosis in Cell Evolution: Life and Its Environment on the Early Earth*. San Francisco: W. H. Freeman.

Matsuda, H., and Y. Harada. 1990. Evolutionarily stable stalk to spore ratio in cellular slime molds and the law of equalization in net incomes. *J. Theor. Biol.* 147: 329–344.

Matthews, S. C., and Missarzhevsky, V. V. 1975. Small shelly fossils of Late Precambrian and Early Cambrian age: A review of recent work. *J. Geol. Soc. London* 131: 289–304.

May, R. M. 1973. *Stability and Complexity in Model Ecosystems*. Princeton, N.J.: Princeton University Press.

Maynard Smith, J. 1966. Sympatric speciation. *American Naturalist* 100: 637–650.

Maynard Smith, J. 1976. Evolution and the theory of games. *American Scientist* 64: 41–45.

Maynard Smith, J. 1978. *The Evolution of Sex*. Cambridge, England: Cambridge University Press.

Maynard Smith, J. 1982. Evolution and the theory of games. Cambridge, England: Cambridge University Press.

Maynard Smith, J., R. Burian, S. Kauffman, P. Alberch, J. Campbell, B. Goodwin, R. Lande, D. Raup, and L. Wolpert. 1985. Developmental constraints and evolution. *Quart. Rev. Biol.* 60(3): 265–287.

Mayr, E. 1963. *Animal Species and Evolution*. Cambridge, Mass.: Belknap Press.

Mayr, E. 1970. Evolution and Verhalten. *Verhandl. Deutschen Zool. Gesell.* 64: 322–336.

Mayr, E. 1974. Behavior programs and evolutionary strategies. *American Scientist* 62: 650–659.

Mayr, E., and W. B. Provine, eds. 1980. *The Evolutionary Synthesis: Perspectives on the Unification of Biology*. Cambridge, Mass.: Harvard University Press.

Michod, R. E. 1982. The theory of kin selection. *Ann. Rev. Ecol. and Syst.* 13: 23–55.

Minelli, A., and S. Bortoletto. 1988. Myriapod metamerism and arthropod segmentation. *Biol. J. Linn. Soc.* 33: 323–343.

Mitchison, T. J. 1988. Microtubule dynamics and kinetochore function in mitosis. *Ann. Rev. Cell Biol.* 4: 527–549.

Mitchison, T. J., and M. W. Kirschner. 1989. Cytoskeletal dynamics and nerve growth. *Neuron* 1: 761–772.

Murray, A. W., and M. W. Kirschner. 1989a. Dominoes and clocks: The union of two views of cell cycle regulation. *Science* 246: 614–621.

Murray, A. W., and M. W. Kirschner. 1989b. Cyclin synthesis drives the early embryonic cell cycle. *Nature* 339: 275–280.

Nakamura, H., and D.D.M. O'Leary. 1989. Inaccuracies in initial growth and arborization of check retinotectal axons followed by course connections and axon remodeling to develop topographic order. *Neuroscience* 9: 3726–3795.

Newgreen, D. F., and C. A. Erickson. 1986. The migration of neural crest cells. *Rev. Cytology* 103: 89–145.

Nijhout, H. F., and D. E. Wheeler. 1982. Juvenile hormone and the physiological basis of insect polymorphisms. *Quart. Rev. Biol.* 572: 109–133.

Noonan, K. M. 1981. Individual strategies of inclusive-fitness-maximizing in *Polistes fuscatus* foundresses. In R. D. Alexander and D. W. Tinkle, eds., *Natural Selection and Social Behavior,* 18–44. New York: Chiron Press.

Nottebohm, F. 1980. Testosterone triggers growth of brain vocal control nuclei in adult female canaries. *Brain Research* 189: 429–436.

Nurse, P. 1990. Universal control mechanics regulating the onset of M-phase. *Nature* 334: 503–508.

Okada, Y. K. 1929. Regeneration and fragmentation in the syllidian polychaetes. *Roux Arch. Entw. Mech. Org.* 115: 542–600.

Okada, Y. K. 1937. La stolonisation et les caractères sexuels du stolon chez les syllidiens polychètes (Études sur les syllidiens III). *Jap. J. Zool.* 7: 441–490.

Otte, D., and J. A. Endler, eds. 1989. *Speciation and Its Consequences*. Sunderland, Mass.: Sinauer Associates.

Page, Jr., R. E., G. E. Robinson, N. W. Calderone, and W. C. Rothenbuhler. 1989. Genetic structure, division of labor, and the evolution of insect societies. In M. O. Breed and R. E. Page, eds., *The Genetics of Social Evolution*. Boulder, Colo.: Westview Press.

Palmer, J. D. 1982. Physical and gene mapping of chloroplast DNA from *Atriplex triangularis and Cucumis sativa. Nucleic Acids Research* 10: 1593–1605.

Pardi, L. 1948. Dominance order in *Polistes* wasps. *Physiol. Zool.* 21(1): 1–13.

Parker, G. A. 1989. Hamilton's rule and conditionality. *Ethol., Ecol. and Evol.* 1: 195–211.

Parker, G. A., R. R. Baker, and V.G.F. Smith. 1972. The origin and evolution of gamete dimorphism and the male-female phenomenon. *J. Theor. Biol.* 36: 529–553.

Parsons, P. A. 1988. Evolutionary rates: Effects of stress upon recombination. *Biol. J. Linn. Soc.* 35: 49–68.

Patel, N. H., T. B. Kornberg, and C. S. Goodman. 1989. Expression of engrailed in grasshopper and crayfish. *Development* 107: 201–212.

Payne, G. S., S. A. Courtneidge, L. B. Crittenden, A. M. Fadly, J. M. Bishop, and H. E. Varmus. 1981. Analysis of avian leukosis virus DNA and RNA in bursal tumors: Viral gene expression is not required for maintenance of the tumor state. *Cell* 23: 311–322.

Peter, M., J. Nakagawa, M. Doree, J. C. Labbe, and E. A. Nigg. 1990. In vitro disassembly of the nuclear lamin and M-phase specific phosphorylation of lamins by cdc2 kinase. *Cell* 61: 591–602.

Peters, N. 1923. Über das Verhältnis der natürlichen zur künstlichen Teilung bei *Ctenodrilus serratus* O. Schmidt. *Zool. Jahrb., Abt. Physiol.* 40: 293–350.

Pimm, S. L., and A. Redfearn. 1988. The variability of population densities. *Nature* 334: 613–614.

Pines, J., and T. Hunt. 1989. Isolation of human cyclin cDNA: Evidence for cyclin mRNA and protein regulation in the cell cycle and for interaction with p34cdc2. *Cell* 58: 833–846.

Presland, R. B., K. Gregg, P. L. Mollog, C. P. Morris, L. A. Crocker, and G. E. Rogers. 1989. Avion keratin genes. I. A molecular analysis of the structure and expression of a group of feather keratin genes. *J. Molec. Biol.* 209: 549–559.

Presland, R. B., L. A. Whitbread, and G. E. Rogers, 1989. Avian keratin genes. II. Chromosomal arrangement and close linkage of three gene families. *J. Molec. Biol.* 209: 561–576.

Provine, W. F. 1971. *The Origins of Theoretical Population Genetics*. Chicago: University of Chicago Press.

Ptashne, M. 1988. How transcriptional activators work. *Nature* 335: 683–689.

Punnett, R. C., ed. 1928. *Scientific Papers of William Bateson*. Cambridge, England: Cambridge University Press.

Raff, R. A., and T. C. Kaufman. 1983. *Embryos, Genes, and Evolution*. New York: Macmillan.

Raup, D. M. 1978. Approaches to the extinction problem. *J. Paleontology* 52: 517–523.

Reznick, D. A., H. Bryga, and J. A. Endler. 1990. Experimentally induced life-history evolution in a natural population. *Nature* 26: 357–359.

Riley, M. A., M. E. Power, and R. C. Lewontin. 1989. Distinguishing the forces controlling genetic variation at the *Xdh* locus in *Drosophila pseudoobscura*. *Genetics* 123: 359–369.

Roeder, S., and G. R. Fink. 1982. Movement of yeast transposable elements by gene conversion. *Proc. Nat. Acad. Sci. USA* 79: 5621–5625.

Rozanov, A. Y. 1986. Problematics of the Early Cambrian. In A. Hoffman and M. H. Nitecki, eds., *Problematic Fossil Taxa,* 87–96. New York: Oxford University Press.

Sadhu, K., S. I. Reed, H. Richardson, and P. Russell. 1990. Human homolog of fission yeast cdc25 mitotic inducer is predominantly expressed in G2. *Proc. Nat. Acad. Sci. USA* 87: 5139–5143.

Sakagami, S. F., and Maeta, Y. 1987a. Sociality, induced and/or natural, in the basically solitary small carpenter bees (Ceratina). In Y. Ito, J. L. Brown, and J. Kikkawa, eds., *Animal Societies: Theories and Facts,* 35–51. Tokyo: Japan Scientific Societies Press, Ltd.

Sakagami, S. F., and Maeta, Y. 1987b. Multifemale nests and rudimentary castes of an "almost" solitary bee *Ceratina flavipes*, with additional observation on multifemale nests of *Ceratina japonica* (Hymenoptera, Apoidea). *Kontyu* 55(3): 391–409.

Sakai, R. K., T. L. Bugawan, G. T. Horn, K. B. Mullis, and H. A. Erlich. 1986. Analysis of enzymatically amplified B-globin and HLA-DQ α DNA with allele-specific oligonucleotide probes. *Nature* 324: 163–166.

Salminen, A., and P. J. Novick. 1987. A ras-like protein is required for post-Golgi event in yeast secretion. *Cell* 49: 527–538.

Salvini-Plawen, L. von. 1985. Early evolution and the primitive groups. In E. R. Trueman and M. R. Clarke, eds., *The Mollusca,* vol. 10, *Evolution,* 59–150. London: Academic Press.

Sander, K. 1988. Studies in insect segmentation: From teratology to phenogenetics. *Development* 104 (Suppl.): 112–121.

Scharloo, W. 1984. Genetics of adaptive reactions. In K. Wöhrmann and V. Loeschcke, eds., *Population Biology and Evolution*, 5–15. New York: Springer-Verlag.

Schmalhausen, I. I. 1949 (1986). *Factors of Evolution: The Theory of Stabilizing Selection*. Chicago: University of Chicago Press.

Seger, J. 1988. Dynamics of some simple host-parasite models with more than two genotypes in each species. *Phil. Trans. Roy. Soc. London* B319: 541–555.

Seilacher, A. 1984. Late Precambrian and Early Cambrian Metazoa: Preservational or real extinctions? In H. D. Holland and A. F. Trendall, eds., *Patterns of Change in Earth Evolution,* 159–168. Heidelberg: Springer-Verlag.

Seilacher, A. 1989. Vendozoa: Organismic construction of the Proterozoic biosphere. *Lethaia* 22: 229–239.

Sepkoski, J. J., Jr. 1981. A factor analytic description of the fossil record. *Paleobiology* 7: 36–53.

Sepkoski, J. J., Jr. 1984. A kinetic model of Phanerozoic taxonomic diversity. III. Post-Paleozoic families and mass extinctions. *Paleobiology* 10: 246–267.

Shackleton, N. J., and N. D. Opdyke. 1973. Oxygen isotope and paleomagnetic stratigraphy of equatorial Pacific core V28–238: Oxygen isotope temperatures and ice volumes on a 10^5-year and 10^6-year scale. *Quaternary Research* 3: 39–55.

Silvertown, J., and D. M. Gordon. 1989. A framework for plant behavior. *Ann. Rev. Ecol. and Syst.* 20: 349–366.

Simpson, G. G. 1944. *Tempo and Mode in Evolution*. New York: Columbia University Press.

Sire, J. V. 1989. Scales in young Polypterius serregatus are elasmoid: New phylogenetic implications. *Am. J. Anatomy* 186: 315–323.

Smith, D. P., B. H. Shieh, and C. S. Zuker. 1990. Isolation and structure of an arrestin gene from Drosophila. *Proc. Nat. Acad. Sci. USA* 87: 1003.

Smith-Gill, S. J. 1983. Developmental plasticity: Developmental conversion versus phenotypic modulation. *American Zoologist* 23: 47–55.

Sober, E. 1984. *The Nature of Selection*. Cambridge, Mass.: MIT Press.

Stavenhagen, J. B., and D. M. Robins. 1988. An ancient provirus has imposed androgen regulation on the adjacent mouse sex-limited protein gene. *Cell* 55: 247–254.

Stanley, S. M. 1979. *Macroevolution: Pattern and Process*. San Francisco: Freeman and Co.

Stanley, S. M. 1990. The general correlation between rate of speciation and rate of extinction: Fortuitous causal linkages. In R. M. Ross and W. D. Allman, eds., *Causes of Evolution, a Paleontological Perspective,* 103–127. Chicago: University of Chicago Press.

Stasek, C. R. 1972. The molluscan framework. *Chemical Zoology* 12: 1–44.

Stewart, M. 1990. Intermediate filaments: Structure, assembly, and molecular interactions. *Current Opin. Cell Biol.* 2: 91–100.

Strand, D. J., and J. F. McDonald. 1985. Copia is transcriptionally responsive to environmental stress. *Nucleic Acid Research* 13: 4401:4410.

Strittmatter, S. M., D. Valenzuela, T. E. Kennedy, E. Neer, and M. C. Fishman. 1990. G0 is a major growth cone protein subject to regulation by GAP-43. *Nature* 344: 836–841.

Stryer, L. 1988. Molecular basis of visual excelation. *Cold Spring Harbor Symp. Quant. Biol.* 53: 283–294.

Suss, E., S. Barash, P. G. Stavenga, H. Stiene, Z. Selinger, and B. Minke. 1989. Chemical excitation and inactivation on photoreceptors of the fly mutants trp and nss. *J. Gen. Physiol.* 94: 965–991.

Thoday, J. M. 1955. Balance, heterozygosity and developmental stability. *Symp. Quant. Biol.* 20: 318–326.

Thomas, K. R., and M. R. Capecchi. 1990. Targeted disruption of the murine int-1 proto-oncogene resulting in severe abnormalities in midbrain and cerebellar development. *Nature* 346: 847–850.

Thomson, K. S. 1988. Morphogenesis and evolution. New York: Oxford University Press.

Tilman, D. 1987. *Resource Competition and Community Structure*. Princeton, N.J.: Princeton, University Press.

Turner, J.R.G. 1977. Butterfly mimicry: The genetical evolution of an adaptation. *Evol. Biol.* 10: 163–206.

Vagvolgyi, J. 1967. On the origin of the molluscs, the coelom, and coelomic segmentation. *Systematic Zoology* 16: 153–168.

Valentine, J. W. 1969. Patterns of taxonomic and ecological structure of the shelf benthos during Phanerozoic times. *Paleontology* 12: 684–709.

Valentine, J. W. 1975. Adaptive strategy and the origin of grades and groundplans. *American Zoologist* 15: 391–404.

Valentine, J. W., ed. 1985. *Phanerozoic Diversity Patterns*. Princeton, N.J.: Princeton University Press.

Valentine, J. W. 1989a. Bilaterians of the Precambrian-Cambrian transition and the annelid-arthropod relationship. *Proc. Nat. Acad. Sci. USA* 86: 2272–2275.

Valentine, J. W. 1989b. Phanerozoic marine faunas and the stability of the earth system. *Palaeogeography, Palaeoclimatology, Palaeoecology* 75: 137–155.

Valentine, J. W. 1990a. Molecules and the early fossil record. *Paleobiology* 16: 94–95.

Valentine, J. W. 1990b. The macroevolution of clade shape. In R. Ross and W. Allmon, eds., *Causes of Evolution: A Paleontological Perspective,* 128–150. Chicago: University of Chicago Press.

Valentine, J. W. 1991. Major factors in the rapidity and extent of the metazoan radiation during the Proterozoic Phanerozoic transition. In A. M. Simonetta and S. Conway Morris, eds., *The Early Evolution of Metazoa and Significance of Problematic Taxa,* 70–73. Cambridge, England: Cambridge University Press.

Valentine, J. W., and D. H. Erwin. 1983. Patterns of diversification of higher taxa: A test of macroevolutionary paradigms. In J. Chaline, ed., Modalité, rythmes et mécanisms de l'evolution biologiques: Gradualism phylétique ou équilibres pontués? *Coll. Int.* 330: 219–223.

Valentine, J. W., B. H. Tiffney, and J. J. Sepkosky, Jr. 1991a. Evolutionary dynamics of plants and animals: A comparative approach. *Palaios* 6: 81–88.

Valentine, J. W., S. M. Awramik, P. W. Signor III, and P. M. Sadler. 1991b. The biological explosion at the Precambrian-Cambrian boundary. *Evol. Biol.* 25: 279–356.

Van Valen, L. 1985. A theory of origination and extinction. *Evol. Theory* 7: 133–142.

Vermeij, G. J. 1987. *Evolution and Escalation*. Princeton, N.J.: Princeton University Press.

Waddington, C. H. 1975. The evolution of an evolutionist. Ithaca, N.Y.: Cornell University Press.

Wagner, G. P. 1989. The origin of morphological characters and the biological basis of homology. *Evolution* 43: 1157–1171.

Wallace, B. 1985. Reflections on the still-"hopeful monster." *Quart. Rev. Biol.* 60: 31–42.

Ward, G. E., and M. W. Kirschner. 1990. Identification of cell cycle-regulated phosphorylation sites on nuclear lamin C. *Cell* 61: 561–577.

Watterson, G. A. 1975. On the number of segregating sites in genetical models without recombination. *Theor. Pop. Biol.* 7: 256–276.

Wcislo, W. T. 1989. Behavioral environments and evolutionary change. *Ann. Rev. Ecol. and Syst.* 20: 137–169.

Weber, J. L., and P. E. May. 1989. Abundant class of human DNA polymorphisms which can be typed using the polymerase chain reaction. *Am. J. Human Gen.* 44: 388–396.

Weber, K., U. Plessmann, H. Dodemont, and K. Kossmagk-Stephen. 1988. Amino acid sequences and homopolymer-forming ability of the intermediate filaments from an invertebrate epithelium. *EMBO Journal* 7: 2995–3001.

Weber, K., U. Plessmann, and W. Ulrich. 1989. Cytoplasmic intermediate filaments of invertebrates are closer to nuclear lamins than are vertebrate intermediate filaments, sequence characterization of two muscle proteins of the nematode. *EMBO Journal* 8: 3221–3227.

Wedel, A., D. S. Weiss, D. Pophom, P. Droge, and S. Kutsu. 1990. A bacterial enhancer functions to tether a transcriptional activator near a promoter. *Science* 248: 486–490.

West-Eberhard, M. J. 1979. Sexual selection, social competition, and evolution. *Proc. of the Zoologist* 51(4): 222–234.

West-Eberhard, M. J. 1983. Sexual selection, social competition, and speciation. *Quar. Rev. Biol.* 58(2): 155–183.

West-Eberhard, M. J. 1984. Sexual selection, competitive communication, and species-specific signals in insects. In L. Trevor, ed., *Insect Communication,* 283–324. London: Academic Press.

West-Eberhard, M. J. 1986. Alternative adaptations, speciation and phylogeny. *Proc. Nat. Acad. Sci. USA* 83: 1388–1392.

West-Eberhard, M. J. 1987. Flexible strategy and social evolution. In Y. Ito, J. L. Brown, and J. Kikkawa, eds., *Animal Societies: Theories and Facts,* 35–51. Tokyo: Japan Scientific Societies Press, Ltd.

West-Eberhard, M. J. 1989. Phenotypic plasticity and the origins of diversity. *Ann. Rev. Ecol. and Syst.* 20: 249–278.

Wheeler, D. E. 1986. Developmental and physiological determinants of caste in social Hymenoptera: Evolutionary implications. *American Naturalist* 128: 13–34.

Whitman, C. O. 1895. *Biological Lectures: The Marine Biological Laboratory at Woods Hole.* Boston: Ginn and Co.

Whittington, H. B. 1974. *Yohoia* Walcott and *Plenocaris* n. gen., arthropods from the Burgess Shale, Middle Cambrian, British Columbia. *Geol. Survey Canada Bull.* 231: 1–21.

Willams, A. F., and A. N. Barclay. 1988. The immunoglobulin superfamily—domains for cell surface recognition. *Ann. Rev. Immunology* 6: 381–405.

Williams, G. C. 1975. *Sex and Evolution.* Princeton, N.J.: Princeton University Press.

Williamson, M. 1981. *Island Populations.* Oxford: Oxford University Press.

Willmer, P. 1990. *Invertebrate Relationships: Patterns in Animal Evolution.* Cambridge, England: Cambridge University Press.

Wilson, E. B. 1925. *The Cell in Development and Heredity.* New York: Macmillan.

Wittenberg, C., K. Sugimoto, and S. I. Reed. 1990. G1-specific cyclins of S. cerevisae: Cell cycle periodically, regulation by mating pheromone, and association with the p34^{cdc28} protein kinase. *Cell* 62: 225.

Wood, W. B., ed. 1988. *The Nematode* Caenorhabditis elegans. Cold Spring Harbor, N.Y.: Cold Spring Harbor Laboratory.

Woodruff, D. S. 1989. Genetic anomalies associated with *Cerion* hybrid zones: The origin and maintenance of new electromorphic variants called hybrizymes. *Biol. J. Linn. Soc.* 36: 281–294.

Wright, S. 1912. Notes on the anatomy of the trematode, *Microphallus opacus*. *Trans. Amer. Micr. Soc.* 31: 167–175.

Wright, S. 1960. The genetics of vital characters of the guinea pig. *J. Cell Comp. Physiol.* 1 (Suppl.): 123–151.

Wright, S. 1982. Character change, speciation, and the higher taxa. *Evolution* 36: 427–443.

Yamamoto, K. R. 1989. *Transcriptional Regulation in Science as a Way of Knowing VI*. Thousand Oaks, Calif.: Cell and Molecular Biology, American Society of Zoologists.

Yokoyama, S., and R. Yokoyama. 1989. Molecular evolution of human visual pigment genes. *Molec. Biol. and Evol.* 6: 186–197.

Young, J. Z. 1974. The central nervous system of Loligo: I. The optic lobe. *Phil. Trans. Roy. Soc. London* B267: 263–301.

Index